Bionanotechnology

Bionanotechnology

Principles and Applications

Anil Kumar Anal

CRC Press
Taylor & Francis Group
Boca Raton London New York

CRC Press is an imprint of the
Taylor & Francis Group, an **informa** business

CRC Press
Taylor & Francis Group
6000 Broken Sound Parkway NW, Suite 300
Boca Raton, FL 33487-2742

First issued in paperback 2020

© 2018 by Taylor & Francis Group, LLC
CRC Press is an imprint of Taylor & Francis Group, an Informa business

No claim to original U.S. Government works

ISBN-13: 978-1-4665-0699-2 (hbk)
ISBN-13: 978-0-367-65640-9 (pbk)

Visit the Taylor & Francis Web site at
http://www.taylorandfrancis.com

and the CRC Press Web site at
http://www.crcpress.com

Contents

Preface

The book *Bionanotechnology: Principles and Applications* deals with a subject area, which is of high interest and importance in all sectors, including biomedical, food, agriculture, and environment. Bionanotechnology combines nanotechnology with biology. Nature has provided us with nanostructures that are extremely efficient and well designed. This book describes the science and technology of controlled building up of new architectures from individual biomolecules and biomacromolecules. Biological systems are essential in nanotechnology and mimicking the natural systems is developing many new applications. This book provides insights into the detail of cellular structures, nanoscale fabrication processes, and their practical applications. The basics of biology and chemistry, with a focus on how to engineer the behavior of molecules at the nanoscale, are also introduced and analyzed. This book is thus designed, so as to (1) focus on the broad accessibility, (2) build design problems of interest that cross the traditional boundaries, (3) accelerate assimilation of new knowledge-spanning multiple domains through individual and construct-centered design problems, and (4) effectively exchange knowledge of state-of-the-art developments and capabilities using collaborative learning projects.

The cells and their entire structures (e.g., cell membranes, lipids, proteins, and nucleic acids) are extremely good at self-assembling complex (e.g., synthesis of proteins and nucleic acids at cellular levels), multifunctional systems at the nanoscale level. By understanding how these systems work, nanotechnologies are developing new biosensing, biomedical, and tissue engineering applications. The use of biological macromolecules as sensors, biomaterials, information storage devices, biomolecular arrays, and molecular machines is significantly increasing. Currently, accumulated knowledge in this area is scattered in few journals and books. This book seeks to bring information of different fields including nanoscale biomaterials; cells; biological macromolecules such as polypeptides, proteins, nucleic acids, lipids, and glycans; interactions between biomaterials and functional bioengineered materials; bionanoencapsulation; controlled release of dosage forms;nanoscale proteomics, genomics, and nanotechnology in immunoisolation and tissue engineering; and so on under one umbrella.

It is essential to understand the fundamental aspects of nanotechnology. This book provides the broader knowledge, including an understanding of biological methods for signal transduction and molecular-recognition systems and how these can be mimicked in biosensing applications. This book attempts to harness various structures, interactions, and functions of biological macromolecules and integrate them with the value addition in multidimensional approaches, including basic structures and interactions of biomacromolecules in developing the biocompatible and ecofriendly devices to be applicable in medicine, agriculture, and food sectors. It encompasses structural biology, biomacromolecular engineering, material science, and extending the horizon of material science.

The aim of this book is to enhance the knowledge to the students, researchers, academicians, professionals, and other stakeholders in the interface between biology, chemistry, and physics; material science; and technology. This book introduces and conveys an understanding about the vast, exciting, and challenging field of bionanoscience, nanotechnology, and the nature behind the developments of these technologies. This book discusses several important and unaddressed aspects of cells, biomacromolecules, and their structures, and interactions and roles in developing the devices for applications, especially in medicine, agriculture, and food sectors. This book also discusses the uptake and health aspects of engineered nanoscale biomaterials and the nanodevices.

Author

Dr. Anil Kumar Anal is the head of the Department of Food Agriculture and Bioresources and an associate professor in Food Engineering and Bioprocess Technology at the Asian Institute of Technology (AIT), Khlong Nung, Thailand. His background expertise is in the food and nutrition security, food safety; food processing and preservation, valorization, as well as bioprocessing of herbs and natural resources, including traditional and fermented foods, microorganisms, agro-industrial waste to fork, and value addition including its application in various food, feed, neutraceuticals, cosmetics, and pharmaceutics. His research interests also include the formulation and delivery of cells and bioactive for human and veterinary applications; controlled release technologies; particulate systems; application of nanotechnology in food, agriculture, and pharmaceutics; and functional foods and food safety. Dr. Anil has authored 5 patents (U.S., World Patents, EU, Canadian, and Indian); more than 100 referred international journal articles; 20 book chapters; 3 edited books; and several international conference proceedings. He has been invited as keynote speaker and expert in various food, biotechnology, agro-industrial processing, veterinary, and life sciences-based conferences and workshops organized by national, regional, and international agencies. Dr. Anil has been serving as advisory member, associate editor, and members of editorial board of various regional and international peer-reviewed journal publications. Dr. Anil has experience of conducting various innovative research and product developments funded by various donor agencies, including the European Union, the FAO, the Ministry of Environment, Japan, and various food and biotech industries.

1

Bionanotechnology and Cellular Biomaterials

1.1 Bionanotechnology

Bionanotechnology is a combination of three terms: *bios* meaning *life*, *nano* (origin in Greek) meaning *dwarf*, and *technologia* (origin in Greek—comprises *techne* meaning *craft* and *logia* meaning *saying*), which is a broad term dealing with the use and knowledge of humanity's tools and crafts. *Biomolecular Nanotechnology* or *Bionanotechnology* is a term coined for the area of study in which nanotechnology has applications in the field of biology, chemistry, and medical sciences. One can also say that *bionanotechnology* is derived by the combination of two terms: (1) *nanotechnology* and (2) *biotechnology*. Bionanotechnology, thus as a subset of nanotechnology implies atomic-level engineering and as a subset of biotechnology and implies atomic-level modification adapting biological machines (Goodsell 2004). The terms *bionanotechnology* and *nanobiotechnology* refer to the intersection of biology and nanotechnology. However, bionanotechnology involves applications of biology to nanotechnology, that is, utilization of biological machines in nanomaterials or nanoscale; for instance; genetic engineering or cellular engineering. Conversely, nanobiotechnology refers to the applications of nanotechnology to study biological system such as nanosensor used for diagnostic purposes or nanoparticles for delivery of active biomolecules (Ramsden 2011).

The word *biotechnology* was first used in early nineteenth century by a Hungarian engineer, Karl Ereky, to refer utilization of the living cells to produce valuable metabolites. The term biotechnology is the combination of Greek words: *bios—life*; *techno—technical*; and *logos—study*. Biotechnology includes a biological process for novel product development and nanotechnology involves engineering and manufacturing at nanometer scales along with atomic precision. Biotechnology can also be termed as, "the application of scientific and engineering principles to the processing of materials by biological agents." Biotechnology is the integration of biochemistry, microbiology, and engineering disciplines to promote industrial applications of microorganisms, engineered cells, tissues, and so on (Amarakoon et al. 2017).

Biotechnology implies the controlled use of biological agents such as cells or cellular components for beneficial use. In general, biotechnology is defined as the utilization of cells, cellular organelles, and living organisms, to produce compounds of interest or the controlled genetic improvement for the benefit of man (Nair 2008).

Along with new discoveries in life-sciences principles and evolution in technologies, biotechnology has undergone various stages of development that can be categorized as ancient, classical, and modern biotechnology. Since ancient period, humans learnt to cultivate and propagated plants, domesticated and interbreed animals to improve their attributes, and got familiar with fermentation and brewing processes. People started using microorganisms to produce wine, beer, cheese, and bread. During the classical biotechnology era, from 1800 to middle of the twentieth century, scientific discoveries and evidence began to outpour. Discoveries made during classical period, form the base for the development of modern biotechnology period. In 1953, Watson and Crick's *Double Helix model of DNA* helped to explain deoxyribonucleic acid (DNA) replication and its role in inheritance. In 1975, Kohler and Milstein postulated the concept of cytoplasmic hybridization and produced the first monoclonal antibodies. Irish scientist Ian Wilmut was successful to clone an adult sheep and developed the first ever cloned animal name *Dolly*. Similarly, there was possibility to sequence the human genome in 2000 AD (Verma et al. 2011). Ancient biotechnology and classical biotechnology also known as first- and second-generation biotechnology were based on technological applications, whereas modern biotechnology, the third generation of biotechnology, is based on the underlying scientific progress (Amarakoon et al. 2017).

Unlike a single discipline, biotechnology is highly multidisciplinary and interdisciplinary related to different scientific disciplines such as pharmaceuticals, immunology, microbiology, genetics, food science and technology, and different aspects of engineering such as mechanical, electronic, food, chemical, and biochemical engineering. Biotechnology has wide applications in the fields of agriculture, livestock, medicine, food, pharmaceutical, aquatic in development of vaccines, drugs, transgenic plants, animals, fish, and so on, and other valuable products and for improving the environment (Bhatia 2005).

Nanotechnology or technology at nanoscale, that is, in the range of 1–100 nm, implies *engineering with atomic precision*. The term *nanotechnology* has been defined in numerous ways, focusing on its design and functionalities. Nanotechnology is the study of material with the size of matter maintained at the nanometer scale to develop innovative devices with distinct properties and various functions. Nanotechnology has also been defined as "the design, synthesis, characterization, and application of materials, devices, and systems that have a functional organization in at least one dimension on the nanometer scale." Figure 1.1 illustrates some of the examples of nanosized biological materials. The benefits associated with nanotechnology

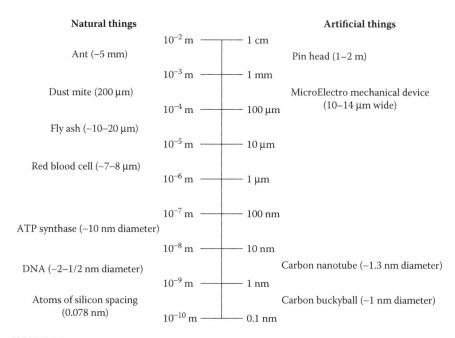

Natural things

Ant (–5 mm)

Dust mite (200 μm)

Fly ash (~10–20 μm)

Red blood cell (~7–8 μm)

ATP synthase (~10 nm diameter)

DNA (~2–1/2 nm diameter)

Atoms of silicon spacing (0.078 nm)

10^{-2} m — 1 cm

10^{-3} m — 1 mm

10^{-4} m — 100 μm

10^{-5} m — 10 μm

10^{-6} m — 1 μm

10^{-7} m — 100 nm

10^{-8} m — 10 nm

10^{-9} m — 1 nm

10^{-10} m — 0.1 nm

Artificial things

Pin head (1–2 m)

MicroElectro mechanical device (10–14 μm wide)

Carbon nanotube (~1.3 nm diameter)

Carbon buckyball (~1 nm diameter)

FIGURE 1.1
Examples of size range of natural and artificial things. (From Picraux, S.T., Nanotechnology, Encyclopaedia Britannica. https://www.britannica.com/technology/nanotechnology, accessed June 30, 2017.)

include formulation of novel product, reduction of energy consumption, and improvement of the functionality (Ramsden 2011).

Nanotechnology has the potential for applications in different sectors such as agriculture, medicine, genetics, food, cosmetics, electronics, and so on. It provides novel techniques, which enhance organoleptic qualities of food, development of new devices as biosensor, packaging materials, and formulation of new encapsulation system for delivery of bioactive compounds (Augustin and Oliver 2012). Various chemicals in the form of nanocapsules, nanocomposites, nanomatrix, and nanoclays are expected to influence the sustainable agriculture development leading to the improvement of the crop yield, reduce chemicals use, and minimize nutrient, water, and soil loss (Iavicoli et al. 2017). Bioactive food ingredients such as essential fatty acids, amino acids, antioxidants, vitamins, and minerals with high biological and functional activities are essential but difficult to be supplied to the body. The problem of poor stability and bioavailability associated with these bioactive food ingredients can be solved by the application of nanotechnology (Jana et al. 2017). Similarly, drugs loaded in nanoparticles have the benefits of delivery and release at the target site, maximum drug action, and minimum side effects, which indicate the role of nanotechnology in pharmaceutical industries (Anal and Stevens 2005; Anal et al. 2006).

Various polymer-based nanoparticles and their conjugates such as carbo-hydrates, proteins, lipids, and nutraceuticals can be utilized as carrier in nanotechnology (Kumar and Smita 2017). Use of antimicrobial components in nanoform along with nanosensor and nanomaterial-based assays in food systems shows the possibilities of future applications of nanotechnology in microbial food safety (Ranadheera et al. 2017).

1.2 Cellular Structures in Bionanotechnology

Cell is the structural unit of all living organisms. The term cell (*Greek, kytos, cell; Latin, cella, hollow space*) was coined in 1665 AD by Robert Hooke, first person to observe the cells in a cork under the primitive microscope. Two German scientists, Schleiden and Schwann, later in 1839 AD outlined the basic features of the cell theory, which describes cell as the basic unit of life. Cell theory has two main components, that is, living things are composed of cells and all cells arise from preexisting cells. Cells vary greatly with respect to shape, for instance, amoebae are irregular in shape, whereas bacteria may exist in rod, spiral, or comma shape; in multicellular organisms, cell shape varies with functions as shown in Figure 1.2. Cell size differs with species; smallest cell (0.2–0.5 µm) found as virus, whereas the largest cell is ostrich egg (6 in. with

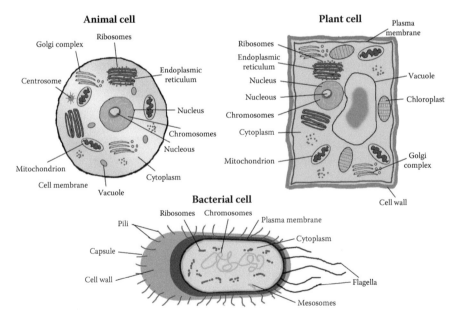

FIGURE 1.2
Structure of bacteria, plant, and animal cells. (From Rogers and Kadner 2017.)

shell and 3 in. without shell). Cell number varies from single cell in unicellular organism to 60,000 billion cells in adult human (Gupta 2008).

Despite differences in external appearance, all living organisms are similar at molecular level, for example, all living organisms store genetic information in nucleic acid, transfer genetic information from DNA and ribonucleic acid (RNA) protein, utilize protein as catalyst, derive energy from adenosine triphosphate (ATP), and possess cell organelles for different functionalities. Living organisms are divided into three divisions: (1) *Bacteria,* (2) *Archaea,* and (3) *Eukarya.*

Bacteria are prokaryotic in nature as they lack membrane bound nucleus, which make them different from eukaryotic plant and animal cells with membrane-bound organelles and genetic materials enclosed by the nuclear membrane. Bacterial cells size vary with species, mostly in the range of 0.5–1 μm in diameter or width; for example, cylindrical typhoid bacteria size ranges from 0.75 to 1.25 μm in width and 2–3 μm in length, or some may be 0.5–2 μm in diameter and more than 100 μm in length. Cell shape of bacteria may be spherical in *cocci*, cylindrical in *bacilli*, or spiral in *spirilla*. Bacterial cells may exist alone or attached to each other in a characteristic arrangement. Bacterial cell size, shape, and arrangement together form the morphology of the cell. Bacteria have external structures such as flagella, which helps to propel them and pili that help bacteria to attach to cell lining the respiratory, intestinal, and genitourinary tracts and also in exchange of genetic materials. Some bacterial cells are surrounded by glycocalyx, which is viscous in nature and helps in capsule formation. On the basis of bacterial cell wall, they are further classified as gram-positive and gram-negative bacteria. Gram-positive bacteria have thicker cell wall (20–25 nm) with large amount of peptidoglycan and teichoic acid as compared to gram-negative bacteria (10–15 nm). Below the cell wall, there is cytoplasmic membrane, which extends in cytoplasm forming tubular structure known as mesosomes that plays role in DNA replication and cell metabolism. However, bacterial cells lack membrane-bound organelles such as mitochondria and chloroplast that are present in eukaryotic cells. Bacterial genetic material, called nucleoid, consisting of a single, circular chromosome is located near center of cells without distinct nuclear membrane but seems to be attached to mesosomes (Pelczar et al. 1986). The structural difference of bacterial cell in comparison to plant and animal cell is illustrated in Figure 1.2.

Animal cells are structurally small with the size in the range of 10–20 μm diameters. They are usually composed of inner cytoplasm and outer double lipid bilayer plasma membrane. Organic and inorganic ions, macromolecules, and cell organelles remain suspended in the cytoplasm, whereas the plasma membrane bounds the cells and internal membrane system divides the cells into compartments. Internal membrane system is formed and supported by filaments and microtubules arranged in network. Cytoskeleton, a microtubule made of protein molecules, functions to support the cell shape and assists in the motility of the cell. The basic form of animal cell constitutes of the cell organelles such as cell membrane, cytoplasm, mitochondria, endoplasmic reticulum, lysosomes, Golgi apparatus, peroxisomes, and nucleus (Young and Francis 2017).

Basically, all cells are composed of biochemical system that are capable to synthesize molecules such as proteins, carbohydrates, lipids, and so on, have outer protective membrane, and possess the ability to store genetic information embedded in nucleic acids. These attributes are similar in both plant and animal cells and both cells possess similar membranes and cell organelles. However, plant cells differ from animal cells in the presence of additional structures: cell wall, chloroplasts, and vacuoles. Cell wall functions to provide rigidity and protection to plant cells. Vacuoles are membrane-enclosed fluid filled organelles found in plant cells including fungi. Vacuoles assist in the storage of nutrients and degradation of unwanted materials inside the cells. And chloroplasts containing chlorophyll have significant role in photosynthesis, which are otherwise absent in animals (Gunning and Steer 1996).

Cytoplasm has complex and subcellular structural suspension known as organelles. Plasma membrane, which surrounds the cellular components forms a protective membrane, which protects the cell from the environment, and regulate the passage of ions and nutrients inside the cell. The adhesion proteins in the plasma membrane help cells to bind specifically to the extracellular matrix. In eukaryotic cells, genetic information is embedded in DNA molecules, which are located inside the nucleus. Nucleus is surrounded by a nuclear membrane, which contains nuclear pore to regulate the passage of molecules into and out of the nucleus. Ribosomes catalyze the protein synthesis via translation using nucleotide sequences of messenger mRNA. Endoplasmic reticulum attached to ribosomes catalyzes the synthesis of cellular lipids, metabolism of drugs, and regulation of cytoplasmic Ca^{+2} concentration. Golgi apparatus functions to process sugar side chains on transmembrane and secreted proteins. Lysosomes secrete hydrolytic enzymes that engulf the foreign materials inside the cell. Mitochondria in the cells secrete a particular enzyme that catalyzes the breakdown of nutrients to synthesize ATP and also respond to toxic stimuli. Peroxisomes also secrete the hydrolytic enzymes to oxidize the intracellular fatty acids. Network of polymer of three protein–actin filaments, intermediate filaments, and microtubules forms the cytoskeleton, which maintains cell shape and assists in cell motility, muscle contraction, transport of organelles, mitosis, and movement of cilia and flagella (Pollard et al. 2017) (Table 1.1).

Nucleic acids are long chains of sugar, phosphate, and nitrogenous base. Based on sugar, they are classified as DNA and RNA, and both of them have gained attention from various disciplines such as biology, physics, and chemistry that signify their role in bionanotechnology. DNA is supercoiled double-stranded nucleic acid, which has a significant role in storage, replication, and realization of genetic information. The size of the human DNA is 2.5 nm in diameter. In 1982, scientist Seeman introduced *structural DNA nanotechnology*, in which synthetic nanoarchitectures were formulated with flexible and artificial DNA with specific sequences with molecular recognition

TABLE 1.1

Dimensions of Subcellular Organelles

Organelles	Dimension ($\times 10^3$ nm)
Peroxisome	0.2–0.7
Primary wall	1–3
Mitochondria	1–10
	Dimension ($\times 10^3$ nm in diameter)
Microbodies	0.1–2
Microtubules	0.5–1
Protein bodies	2–5
Chloroplast	4–6
Nucleus	5–10
Plasmodesmata	2–40
Nuclear envelope pores	30–100
Golgi apparatus	0.9 (Individual cisternae)
	50–280 (coated vesicles)

Source: Dashek, W.V., An introduction to cells and their organelles, in *Plant Cells and Their Organelles*, Dashek, W.V. and Miglani, G.S. (Eds.), pp. 1–24, Chichester, UK: John Wiley & Sons, 2017.

properties. Such models have been used as synthons in structural DNA technology because of the ability of DNA to exhibit specific base pairings. DNA double-crossover molecules (DX molecules) have been utilized as building blocks for assembly of two-dimensional (2D) and three-dimensional (3D) DNA crystals. Similarly, RNA molecules that have major function as carrier of genetic information, genetic code, and amino acids during protein synthesis have also been utilized in a similar way to design various nanostructures. RNA nanostructures were constructed successfully in 1998 through the self-assembly of reengineered natural RNA. Nucleic acid has been utilized to fabricate nanobiomaterials utilizing polymers, metals, or semiconductor nanoparticles (Kundu et al. 2017).

Deoxyribonucleic acid nanotechnology is progressing continuously and is expected to be more applicable in medical science. DNA has the potentials to be used as drug carriers and vaccine adjuvants as well as for diagnostic purposes. The DNA origami nanotubes have been used as molecular rulers for the calibration of super-resolution microscopy. These nanostructures have also been used to design nanoelectronic devices (Liu and Ellington 2014).

Broadly, biomaterials are defined as any natural or synthetic materials that come in contact with biological system for their specific functionalities. Biomaterials have basic functions in medical science where it is originally defined as "nonviable material used in a medical device, intended to interact with biological system" and the interaction may be direct or indirect. Biomaterial field is highly interdisciplinary; its definition has undergone several modifications along with the development of biomaterials, their intended function along with improvements in physics, biology, chemistry, material science, engineering, biotechnology, and medicine. Initially, only synthetic materials were regarded as biomaterials, whereas natural materials were included in a definition later. Biomaterials intended for medical purpose should be biocompatible, that is, it should not induce toxicity, lack foreign body, and promote normal healing. Biomaterials may exist in solid, liquid, or gel forms and could be synthetic, naturally derived, or semisynthetic (Kulinets 2015).

Some of the commonly utilized synthetic polymers at nanoscale include polylactic acid (PLA), polyglycolic acid (PGA), and their copolymer, poly(lactic-co-glycolic acid) (PLGA). Clinically, these polymers have been utilized to prepare urethral tissue and substitute bladder in patients. Physical properties of polymers such as mechanical strength, tensile strength, and degradation resistance can be tailored made based on the intended purpose, which makes synthetic polymer a suitable substitute in the medical field (Chen and Liu 2016). Natural biomaterials are similar to native tissue intended to be engineered and possess properties that can assist in reconstruction, repair, and regeneration. This makes natural biopolymer an important subset to be used as biomaterials in tissue engineering. The common natural biopolymers utilized as biomaterials include proteins, polysaccharides, and polynucleotides (Kulinets 2015).

Proteins, nucleic acids, polysaccharides, and lipids are four basic and functional biopolymers utilized by most of the cells for the cellular activities. Proteins are polymers of amino acids linked together by an amide linkage (peptide bonds) between the amine of one amino acid and carboxyl group of adjacent amino acid. The proteins with rigid amide bonds have been utilized to build nanomachines, nanostructures, and nanosensors. Nucleic acids are specialized for application in nanoscale information storage and retrieval. Common natural lipids, phospholipids, and glycolipids are composed of both polar and nonpolar group because of which they are amphiphilic in nature such as proteins. Polysaccharides are heterogeneous polymer of sugars linked by hydrogen bonding between the hydroxyl groups and are responsible for cellular structure and energy storage. Lipids are the main components of the cell membranes, which are impermeable to ions and larger polar molecules (Goodsell 2004).

In medical field, bionanomaterials are utilized as drug delivery carrier, coating materials, membranes, cell culture, and healing materials. Use of hybrid biomaterials integrated with dermal and epidermal component have been developed *in vitro* for skin tissue engineering (Greenwood 2016). Carboxymethyl chitosan with an improved solubility was proved to be a suitable substitute for producing drug-loaded biomaterials and for designing nanotechnology-based systems (Fonseca-Santos and Chorilli 2017). Orthopedic infections and difficulties during treatment have been tackled by the use of suitable biomaterials such as antibiotic-loaded collagen fleeces and ceramic or composite calcium-based bone graft substitutes such as calcium sulfates, calcium phosphates, and hydroxyapatite (van Vugut et al. 2017). Biomaterials are the fundamental components of tissue engineering, which aims to replace diseased, damaged, or missing tissues with reconstructed functional tissue. Suitable biomaterials have been utilized in tissue engineering that assists in stem cell differentiation and maintenance of phenotypic characteristics (Keane and Badylak 2014; Gajendiran et al. 2017).

References

Amarakoon, I. I., C. L. Hamilton, S. A. Mitchell, P. F. Tennat, and M. E. Roye. 2017. Biotechnology. In *Pharmacognosy* (Eds.) S. Badal and R. Delgoda, pp. 549–563. Tokyo, Japan: Academic press.

Anal, A. K. and W. F. Stevens. 2005. Chitosan-alginate multilayer beads for controlled release of ampicillin. *International Journal of Pharmaceutics* 290:45–54.

Anal, A. K., W. F. Stevens, and C. Remunan-Lopez. 2006. Ionotropic cross-linked chitosan microspheres for controlled release of ampicillin. *International Journal of Pharmaceutics* 312:166–173.

Augustin, M. A. and C. M. Oliver. 2012. An overview of the development and applications of nanoscale materials in the food industry. In *Nanotechnology in the Food, Beverage and Nutraceutical Industries* (Eds.) Q. Huang. New Delhi, India: Woodhead Publishing.

Bhatia, S. C. 2005. *Textbook of Biotechnology*. New Delhi, India: Atlantic Publishers and Distributors.

Chen, F. M. and X. Liu. 2016. Advancing biomaterial of human origin for tissue engineering. *Progress in Polymer Science* 53:86–168.

Dashek, W. V. 2017. An introduction to cells and their organelles. In *Plant Cells and Their Organelles* (Eds.) W. V. Dashek and G. S. Miglani, pp. 1–24. Chichester, UK: John Wiley & Sons.

Fonseca-Santos, B. and M. Chorilli. 2017. An overview of carboxymethyl derivatives of chitosan: Their use as biomaterials and drug delivery system. *Material Science and Engineering* 77:1349–1362.

Gajendiran, M., J. Choi, S. J. Kim, K. Kim, H. Shin, H. J. Koo, and K. Kim. 2017. Conductive biomaterials for tissue engineering applications. *Journal of Industrial and Engineering Chemistry* 51:12–26.

Goodsell, D. S. 2004. *Bionanotechnology Lessons from Nature*. Hoboken, NJ: Wiley-Liss.

Greenwood, J. E. 2016. Hybrid biomaterials for skin tissue engineering. In *Skin Tissue Engineering and Regenerative Medicine* (Eds.) M. Z. Albanna and J. H. Holmes IV, pp. 185–210. Amsterdam, the Netherlands: Academic Press.

Gunning, B. E. S. and M. W. Steer. 1996. *Plant Cell Biology: Structure and Function. Life Science Series*. Boston, MA: Jones and Bartlett Publishers.

Gupta, P. K. 2008. *Molecular Band Genetic Engineering*. New Delhi, India: Global Media Publication.

Iavicoli, I., V. Leso, D. H. Beezhold and A. A. Shvedova. 2017. Nanotechnology in agriculture: Opportunities, toxicological implications and occupation risks. *Toxicology and Applied Pharmacology* 329:96–11.

Jana, S., A. Gandhi and S. Jana. 2017. Nanotechnology in bioactive food ingredients: Its pharmaceutical and biomedical approaches. In *Nanotechnology Applications in Food* (Eds.) A. E. Oprea and A. M. Grumezescu, pp. 21–41. Tokyo, Japan: Academic Press.

Keane, T. J. and S. F. Badylak. 2014. Biomaterials for tissue engineering applications. *Seminars in Pediatric Surgery* 23:112–118.

Kulinets, I. 2015. Biomaterials and their applications in medicine. In *Regulatory Affairs for Biomaterials and Medical Devices* (Eds.) S. F. Amato and R. M. Ezzell Jr, pp. 1–10. Amsterdam, the Netherlands: Woodhead Publishing.

Kumar, B. and K. Smita. 2017. Scope of nanotechnology in nutraceuticals. In *Nanotechnology Applications in Food* (Eds.) A. E. Opera and A. M. Grumezescu, pp. 43–63. Tokyo, Japan: Academic Press.

Kundu, A. S. N. and A. K. Nandi. 2017. Nucleic acid based polymer and nanoparticle conjugates: Synthesis, properties and applications. *Progress in Material Science* 88:136–185.

Liu, C. and A. D. Ellington. 2014. DNA nanotechnology: From biology and beyond BT. In *Nucleic Acid Nanotechnology* (Eds.) J. Kjems, E. Ferapontova, and K. V. Gothelf, pp. 135–169. Heidelberg, Germany: Springer.

Nair, A. J. 2008. *Introduction to Biotechnology and Genetic Engineering.* New Delhi, India: Infinity Science Press LLC.

Picraux, S. T. 2017. Nanotechnology, Encyclopaedia Britannica. https://www.britannica.com/technology/nanotechnology. (accessed June 30, 2017).

Pelczar, M. J., E. C. S. Chan, and N. R. Krieng. 1986. *Microbiology: International Student ed.* New York: McGraw-Hill.

Pollard, T. D., W. C. Earnshaw, J. Lippincott-Schwartz, and G. T. Johnson. 2017. *Cell biology (Third edition).* Amsterdam, Netherlands, Elsevier.

Ramsden J. J. 2011. *Nanotechnology.* Tokyo, Japan: William Andrew Publishers.

Ranadheera, C. S., P. H. P. Prasanna, J. K. Vidanarachchi, R. McConchie, N. Naumovski, and D. Mellor. 2017. Nanotechnology in microbial food safety. In *Nanotechnology Application in Food* (Eds.) A. E. Opera and A. M. Grumezescu, pp. 245–65. Tokyo, Japan: Academic Press.

Rogers, K. and R. J. Kadner. 2017. Bacteria. Encyclopædia Britannica, inc. https://www.britannica.com/science/bacteria. (accessed June 13, 2017).

Verma, A. S., S. Agrahari, S. Rastogi, and A. Singh. 2011. Biotechnology in the realm of history. *Journal of Pharmacy and Bioallied Sciences* 3:321–323.

van Vugut, T. A., J. A. P. Geurts, J. J. Arts, and N. C. Lindfors. 2017. Biomaterials in treatment of orthopedic infections. In *Management of Periprosthetic Joint Injections* (Eds.) J. J. Chris and J. Geurts, pp. 41–68. Amsterdam, the Netherlands: Woodhead Publishing.

Young, R. and S. Francis. 2017. Form and function of the animal cell. In *Pharmacognosy* (Eds.) S. Badal and R. Delgoda, pp. 459–75. Tokyo, Japan: Academic Press.

2

Nanostructured Cellular Biomolecules and Their Transformation in Context of Bionanotechnology

2.1 Nucleic Acids

Both nucleic acids: deoxyribonucleic acids (DNA) and ribonucleic acids (RNA), with significant role in storage, replication, and realization of genetic information have gained attention from various disciplines such as biology, physics, and chemistry that signify their roles in bionanotechnology. A novel research field, *structural DNA nanotechnology* began to emerge in early 1980s, which developed nanoarchitectures with flexible and specific sequence artificial DNA as molecular recognition system. The DNA structures are inherently at nanoscale; for instance, DNA diameter is about 2 nm (20 Å), the helical periodicity is around 3.5 nm per turn and the base separation distance is around 3.4 Å. These properties of DNA have been utilized for the fabrication of DNA-based nanodevices (Seeman 2007). Similarly, RNA molecules that have major function as carrier of genetic information, code, and amino acids during protein synthesis, have also been utilized to design various nanostructures. First, RNA nanostructure was constructed success-fully in 1998 through the self-assembly of reengineered natural RNA. Since then, nucleic acids have been utilized to fabricate nanobiomaterials utilizing polymers, metals, or semiconductor nanoparticles (Guo 2010).

Bionanotechnology is progressing continuously and is expected to be more applicable in medical sciences. Various nanostructures, including the DNA, and other bioinspired nanomaterials have potential use as drug carriers and vaccine adjuvants. DNA nanostructures are also applicable for analysis and diagnostic purposes. DNA origami nanotubes have been used as molecular rulers for the calibration of super-resolution microscopy. These nanostructure-based materials are used to design nanoelectronic devices (Liu and Ellington 2014).

Structurally, nucleic acids are the polymers of nucleotides held by 5′ and 3′ phosphate bridges and constituted as major organic matters of the nuclei of biological cells. Nucleotides are composed of nitrogenous bases, pentose sugars, and phosphate groups. Nitrogenous bases are heterocyclic with rings containing carbon and nitrogen, which are further divided into two groups: (1) purine, with a pair of fused rings and (2) pyrimidines, with single ring. Adenine (A) and guanine (G) are purines and thymine (T), uracil (U), and cytosine (C) are pyrimidines. The sugar component can be ribose or deoxy-ribose. Depending on the presence of types of sugar molecules, nucleic acids are further classified into two classes: (1) DNA containing 2-deoxy-D-ribose and (2) RNA containing D-ribose sugar (Wong and Jameson 2011).

The nucleic acids, DNA and RNA, each are made up of four different nucleotides, each with a common structure: a phosphate group linked to pentose sugar by phosphodiester bond, which in turn is linked to nitrogenous base. In DNA, bases are adenine (A), guanine (G), thymine (T), and cytosine (C), whereas in RNA, all bases are same as DNA except thymine being replaced by uracil (U). Nucleic acids are chains of nucleotides such that C-3′ atom of sugar of one nucleotide is linked to C-5′ atom of the neighboring sugar of another nucleotide by a phosphodiester bond and the nitrogenous base attached to C-1′ end of sugar by a β-glycosidic bond (Ullmann 2007).

DNA is the chemical basis of heredity and generally regarded as the reserve bank of genetic information. DNA is exclusively responsible to maintain the identity of different species of organisms. Watson and Crick (1953) proposed the double helical structure of DNA, the model's four major following features remain the same. DNA is a double-stranded helix, with the two strands connected by hydrogen bonds. Bases adenines (A) are always paired with thymines (T), and cytosines (C) are always paired with guanines (G), which is consistent with and accounts for Chargaff's rule. The Chargaff's rule suggests that the Adenine (A) is equal to Thymine (T) and Cytosine (C) is equal to Guanine (G) in double-helix DNA molecules. Most of the DNA double helices are right handed. Only one type of DNA, called Z-DNA, is left handed. The DNA double helix is antiparallel, which means that the 5′ end of one strand is paired with the 3′ end of its complementary strand (and vice versa). As shown in Figure 2.1, nucleotides are linked to each other by their phosphate groups, which bind the 3′ end of one sugar to the 5′ end of the next sugar. The double-stranded DNA structure is greatly stabilized because of the presence of thousands of hydrogen bonds, hydration of the phosphate group, and hydrophobic interactions between the aromatic ring systems resulting in the stacking of bases (Lodish et al. 2000).

The three-dimensional (3D) double-helical structure of DNA (Figure 2.1) comprises two polynucleotide chains each known as DNA strand. In a DNA strand, nucleotides are covalently linked together through alternating sugar–phosphate group that forms the backbone located outside of the molecule, whereas two polynucleotide chains are held together with hydrogen bonds

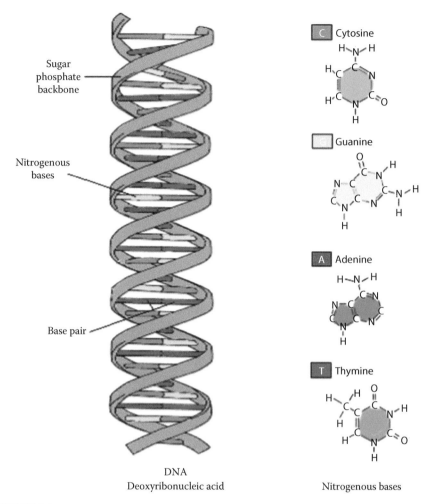

FIGURE 2.1
Watson–Crick model for the structure of DNA. (From Nelson, D.L. and Cox, M.M., *Lehninger Principles of Biochemistry*, W.H. Freeman, New York, 2012. With Permission.)

between the bases, which are located inside the helix (Alberts et al. 2002). Not only are the DNA base pairs connected through hydrogen bonding, but the outer edges of the nitrogen-containing bases are exposed and available for potential hydrogen bonding as well. These hydrogen bonds provide easy access to the DNA for other molecules, including the proteins that play vital roles in the replication and expression of DNA.

The RNA structure is primarily similar to DNA with 3,5-phosphodiester bonds as backbone of the molecule and exists as single strand; however, RNA has a ribose sugar and uracil as pyrimidine in place of thymine. The hydroxyl group present in C_2 position of ribose (Figure 2.2) increases the

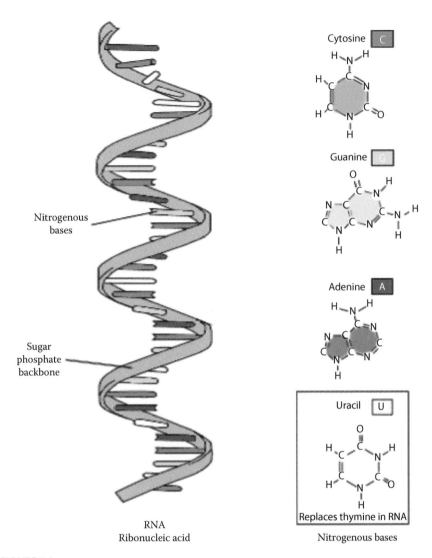

Cytosine C

Guanine G

Adenine A

Uracil U

Replaces thymine in RNA

Nitrogenous
bases

Sugar
phosphate
backbone

RNA
Ribonucleic acid

Nitrogenous bases

FIGURE 2.2
Structure of ribonucleic acid. (From Nelson, D.L. and Cox, M.M, *Lehninger Principles of Biochemistry*, W.H. Freeman, New York, 2012. With Permission.)

chemical stability of RNA and it can be cleaved into mononucleotides by an alkaline reagent, which is not possible in DNA. RNA is an unbranched single-stranded molecule that can exist in different structures by pairing of complementary bases within RNA molecule as shown in Figure 2.2. Secondary structures such as *hairpins* are formed by base pairing within approximately 5–10 nucleotide and *stem loops* structures are formed by base pairing separated by around 500 to several thousand nucleotides. Tertiary structure of RNA, *pseudoknot*, is formed by folding back in a hairpin and formation of a second stem—loop structure. The RNA molecules with different conformations and sizes are assigned for different functions in the cell (Lodish et al. 2000). There are three different types of functional RNAs: (1) messenger RNA (mRNA), (2) ribosomal RNA (rRNA), and (3) transfer RNA (tRNA). The small nuclear RNA exists only in eukaryotic cells and is responsible for guiding the formation of mRNA by splicing of pre-mRNA (Alberts et al. 2002).

1. *Ribosomal ribonucleic acid*: About two-thirds of ribosomal mass is composed of rRNA, which makes up almost all key sites of ribosomal function. Ribosomes, therefore are ribonucleoprotein, which are often regarded as RNA enzymes or ribozymes. Prokaryotic ribosomes are composed of 30S (small subunit) and 50S (large subunit) The 30S subunit consist of 16S rRNA (RNA molecules with more than 1500 nucleotides) and 21 proteins (referred as S1 through S21), whereas 50S subunit is composed of 23S (RNA molecules with around 2900 nucleotides) and the 5S (RNAs with 120 nucleotides) along with 34 different proteins (referred as L1 through L34). Eukaryotic ribosome, which is slightly larger than prokaryotic ribosome, is composed of 40S small subunit and 60S large subunit. rRNA functions to read the genetic code of mRNA and synthesizes the specific proteins to that particular code. The rRNA of small subunit functions to decode the genetic information and the large subunit functions to add the amino acid onto the growing peptide chain during the protein synthesis (Swayze et al. 2007).

2. *Messenger ribonucleic acid*: mRNA molecules are the RNA molecules, which carry the genetic information for specific amino acid sequence of proteins. In prokaryotic cells, mRNA is synthesized on the DNA template by single RNA polymerase and is composed of either monocistronic or polycistronic molecules. While in eukaryotic cells, mRNA is composed of monocistronic molecules. mRNA molecule folds into 3D structures by base pairing forming different structures such as stem and internal loops, bulges, junction, and pseudoknots (Swayze et al. 2007).

3. *Transfer ribonucleic acid*: tRNA is the key to understand the genetic information of mRNA. Each amino acid has specific tRNA to carry it to rRNA and transfer it to the growing polypeptide chain based on the genetic code in mRNA. The tRNA molecule consists

of three base sequences, which result in the correct attachment of amino acids (Lodish et al. 2000). tRNA is composed of around 75–90 nucleotides base pairing resulting in a secondary structure with a stem and three loops similar to that of a cloverleaf (Ullmann 2007).

2.2 DNA Replication and Genetic Transformation

The genetic information in terms of base sequence in a double-stranded DNA (single or double stranded DNA or RNA in viruses) needs to be transferred with accuracy. DNA replication is a process with perfected proofreading and repair mechanisms with chance of one mismatch for every 108–1011 bases in a new DNA. Figure 2.3 illustrates the schematic details of semiconservative modules of DNA replication.

The replication of DNA has three distinct stages that include initiation, elongation, and termination. Initially, helicase enzyme untwists the helical structure, breaks the hydrogen bond between base pairs, separates double strand into single DNA strand, and creates fork-like structures known as replication fork at the site known as *origin of replication*. Primer, a short

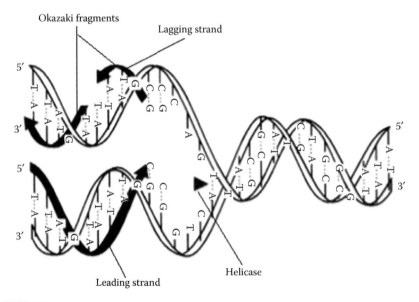

FIGURE 2.3
Semiconservative model of DNA replication. (From Ullmann, F., Nucleic acids genetic engineering is a separate keyword, in *Ullmann's Biotechnology and Biochemical Engineering* (Ed.), Wiley-VCH Staff, Wiley-VCH, Weinheim, Germany, pp. 1657–2187, 2007. With Permission.)

double-stranded piece of nucleic acid synthesized by RNA polymerase, binds to the site of action. DNA polymerase III synthesizes a new DNA from deoxyribonucleotide triphosphates with the release of pyrophosphate; the cleavage of pyrophosphate by a pyrophosphatase provides the energy required for DNA biosynthesis. The addition of new nucleotide always takes place at the 3–OH group of the deoxyribose sugar; therefore, all biologically synthesized nucleic acids grow in the 5′ → 3′ direction. As DNA strands are antiparallel in 3′ → 5′ strand, synthesis occurs backward and discontinuously forming Okazaki fragments, which are later joined by DNA ligase. Thus, the formed strand is known as lagging strand, whereas the strand synthesized continuously in forward direction is known as the leading strand. Addition of nucleotides occurs at a fast pace, around 750 bases per second at each fork. As replication proceeds, the duplicated strand loops down. When the fork forms a full circle and meets at its ends, ligases move along the lagging strand to link the fragments and separate the circular daughter molecules. DNA polymerase I removes the primers and replaces with the DNA. The replication of DNA is a semiconservative process, that is, one strand of each of the two new daughter molecules of DNA is an old strand and the other one is a newly synthesized one (Ullmann 2007).

2.3 Central Dogma: Gene–Protein Connections

The sequence of nitrogenous base in DNA carries the genetic information in the form of genes. Three successive nucleotides, known as codons, act as a code for specific amino acid, for example, sequence of GCT, GCA, GCG, or GCC code for alanine. Out of 64 possible codons, 61 code for amino acids, whereas 3 serve as specific stop signal. Gene expression is therefore based on the recognition of nucleic acid sequence. The genetic code is universal, which implies that prokaryotes can synthesize eukaryotic proteins from eukaryotic DNA, which forms the foundation of genetic engineering.

DNA is organized into genes that regulate protein synthesis utilizing RNA as a mediator. The genetic information in DNA acts as a template for mRNA synthesis, which further acts as a template for protein synthesis. This interrelationship between DNA, RNA, and protein is known as central dogma in molecular biology.

2.3.1 Transcription

Transcription is the process by which genetic information is transferred from DNA to mRNA synthesized from ribonucleoside triphosphates utilizing the energy released by the removal of pyrophosphate. The site for initiation of transcription process is known as promoters, whereas in bacteria it is known as operon from which several genes are cotranscribed from same promoters. DNA is double stranded in antiparallel orientation, that is, top strand runs from 5′ → 3′ direction, whereas the bottom strand runs from the 3′ → 5′ direction. During the transcription process, RNA synthesis proceeds in 5′ → 3′ direction such that the bottom strand is known as a template strand as it acts as a template for RNA synthesis, whereas the top strand is known as the coding strand as its sequence corresponds to the DNA version of the mRNA. Top strand is also called sense strand as mRNA with this sequence will make the correct protein, whereas the bottom strand is known as the antisense strand. The transcription process initiates with the unwinding of DNA helix and synthesis of short chain RNA primer, process being catalyzed by RNA polymerase. The chain length elongates with the addition of ribonucleotides triphosphates forming phosphodiester linkage. The transcription process terminates as it reaches to a specific gene sequence known as termination sequence or binding of a specific protein *rho* that causes disassembly of template, enzyme, and synthesized mRNA (Moran et al. 2012).

2.3.2 Translation

Translation is a very intricate multistep irreversible process occurring in the translation complex. The translation complex is the assembly of four different components: (1) two ribosomal subunits, (2) more than 50 protein factors known as initiation factors, (3) mRNA, and (4) aminoacyl-tRNA. Ribosome catalyzes the peptide bond formation, protein factors assist ribosome in each steps of translation, mRNA carries the genetic information from DNA specifying the protein's sequence, and the aminoacyl-tRNAs carry activated amino acids for protein synthesis. Ribosomes are composed of rRNA and protein synthesis takes place in ribosomes. The ribosomes are composed of two subunits: 30S and 50S subunit, which combine to form active 70S in prokaryotes and 40S and 60S subunit, which combine to form active 80S ribosome in eukaryotes. The subunits are separated by a narrow neck and the protrusion extending from the base of ribosome forms a cleft where the mRNA molecule attaches. Ribosomes have two sites for binding of aminoacyl-tRNA such that their anticodons interact with the correct mRNA codons. The aminoacylated ends of two tRNAs are positioned at the site of peptide bond formation. During protein synthesis, ribosome has a role to hold mRNA and growing polypeptide chains and to accommodate the several protein factors.

Protein synthesis initiates with the assembly of the translation complex. Ribosome identifies the initiation codon, which is usually AUG (codes for methionine), but GUG, UUG, or AUU may also be used. Among the two methionyl-tRNA[Met] molecule that translates AUG codons, initiator tRNA is used at the initiation codon. At the end of the initiation step, mRNA is positioned in such a way that the next codon can be translated during the elongation step, the initiator tRNA is attached to P site and A site of ribosome is ready to receive incoming aminoacyl-tRNA. During chain elongation process, tRNA with the correct aminoacyl-tRNA attach into the A site. The peptidyl transferase catalyzes a transfer of the amino acid from the P site to the amino acid at the A site with the formation of peptide bond such that the growing polypeptide chain is covalently attached to the tRNA in the A site, forming a peptidyl-tRNA. The first amino acid on the polypeptide has a free amino group, so it is called the *N*-terminal and the last amino acid has a free carboxylic group, so it is called the *C*-terminal. The mRNA shifts by one codon such that two tRNAs at the P and A site translocate. Following this, deaminoacylated tRNA is displaced from P to exit, E site and peptidyl-tRNA is displaced from A to P site. When one of the three stop codons (UAA, UAG, UGA) on the mRNA is reached, tRNA does not recognize these termination codons and release factor causing the hydrolysis of the peptidyl-tRNA releasing polypeptide chain (Moran et al. 2012).

2.3.3 Transformation

The term transformation in bionanotechnology is defined as the alteration of genotype of a cell due to incorporation of the external genetic material (DNA). The introduction of new genetic material can be done in nonbacterial cells, plants, and animals cells. The process of introducing new genetic material in animal cells is known as transfection. Transformation was first discovered in 1928 by Frederick Griffith in *Streptococcus pneumoniae*, who demonstrated that nonvirulent strain of *Streptococcus pneumoniae* could be virulent on exposure to heat-killed strains (Griffiths et al. 2000). Transformation, however, is different from other horizontal gene-transfer process: conjugation (involves direct contact between two cells) and transduction (gene transfer involving bacteriophage). Transformation can occur either naturally or is induced artificially. Natural transformation implies active uptake of free (extracellular) DNA (plasmid and chromosomal) and the heritable incorporation of its genetic information, whereas artificial transformation is a process in which DNA is introduced into the bacterial cells *in vitro*. The transformation process for gene transfer does not need active donor but the recipient cells need to be metabolically active to be able to take up DNA. Further, donor and recipient cells do not need to be genetically related. The introduction of foreign DNA can be done by treatment of cells with various chemical reagents such as chlorides ($CaCl_2$, $MgCl_2$), chelating agents such as ethylene diamine tetraacetic acid (EDTA), or enzymes (muraminidases or peptidase) to form protoplasts. It can also be done by various physical treatments such as fusion of cells or protoplasts with DNA with cells or with DNA packaged in liposomes, freezing and thawing of cells, exposure of cells to electric fields (electroporation), and the bombardment of cells with small particles and transporting DNA into the cytoplasm (Lorenz and Wackernagel 1994).

   Transformation plays a significant role in adaptation and ecological diversification of the organisms. Several bacteria and Archaea acquire foreign genetic materials by conjugation, transduction, and natural transformation.

In contrast to conjugation and transduction, during transformation, the recipient cell itself initiates the transfer of exogenous genetic material. While conjugation and transduction rely on extrachromosomal genetic elements promoting their own maintenance and distribution, natural transformation is a part of the normal physiology of the competent bacterium and is therefore uniquely adapted to the needs of the host. Transformation has been extensively studied in bacterial species such as *Streptococcus pneumoniae*, *Neisseria* spp., *Bacillus subtilis*, and *Haemophilus influenzae* (Johnsburg et al. 2007).

2.4 Hybridization

Hybridization refers to the method in which molecules of nucleic acid (either single-stranded DNA or RNA) are bound to the complementary sequence, that is, adenine (A) pairs with thymine (T) or uracil (U) or vice versa and guanine (G) pairs with cytosine (C) or vice versa. Blotting is an important technique to study the hybridization in nucleic acid. The chemical basis for nucleic acid hybridization relies in the reversible helix-coil transition of the nucleic acid, which can associate as a double-stranded or dissociate into a single-stranded polymer based on the physicochemical condition of the surrounding such as temperature, ionic strength, and presence of denaturing agents. Disassociated single-stranded nucleic acid can anneal to complementary sequence forming homologous DNA (Edberg 1985).

Hybridization techniques display a wide scope of application, including DNA–DNA hybridization, polymerase chain reaction (PCR), southern blots, northern blots, forensic DNA testing, and in medical science for diagnostic purpose. DNA–DNA hybridization is a method utilized to measure the extent of genetic similarity between two organisms. For this, labeled DNA sequence of known organisms is incubated with an unknown DNA sequence to allow formation of hybrid double-stranded DNA. Hybridized sequence with great similarity will bind strongly and dissociates only at higher temperatures compared to dissimilar sequence. This method is used as a taxonomic gold standard for species delineation in Archaea and bacteria. DNA–DNA hybridization similarity below 70% signifies distinct relation between organisms. However, this process being laborious and tedious with high chance of error, alternative methods are being researched (Meier-Kolthoff et al. 2013). Fluorescence *in situ* hybridization is a method that utilizes fluorescent probes to detect specific DNA sequence. This technique has wide application for identification of species, gene, RNA (mRNA), tumor, or cancerous cells. High-sensitivity detection, simultaneous assay of multiple species, automated data collection, and analysis have made this method a foremost biological assay (Levsky and Singer 2003).

2.5 Proteins and Peptides

Peptides have been utilized as building units in nanotechnology because of their wide applications for synthesis of materials and fabrication of nanodevices. Specific properties such as superior specificity for target binding in antibodies and biological recognition function facilitate utilization of peptides and proteins in the development of nanomachines with specific recognition function. Similarly, devices fabricated with collagen triple helix peptides can undergo structural change in response to pH, ionic strength, temperature, and electric/magnetic fields. Protein structures with catalytic function can be utilized as a template for 3D crystallization process, which would enable the growth of various materials in aqueous condition at low temperature. Growth of biomimetic nanomaterials is useful for therapeutic and diagnostic purposes. Further, peptides can be utilized as signal transduction pathways in cells such as neuropeptides used as neurotransmitters. These properties of peptides and protein-related materials make them a possible unit in the design of nanodevices such as nanoreactors, nanobiosensor, nanochips, and stimulus-responsive materials. This overall helps to widen the scope of bionanotechnology (De La Rica and Matsui 2010).

Proteins, polymers of amino acids, are essential part of the living system. Proteins serve as important biological roles; proteins act as biocatalyst such as enzymes, bind with other molecules, and help in the storage and transportation such as hemoglobin binds and transports O_2 and CO_2, serve as pores and channels in membranes, provide shape and support to cells, assist in mechanical function of living organisms such as helping in muscles' contraction, movement of flagella, gene expression, regulation of biochemical activities, acting as receptors for various ligands, and some proteins have highly specialized functions such as antibodies (Moran et al. 2012).

Protein molecules have up to four levels of orientation known as primary, secondary, tertiary, and quaternary structures (Figure 2.4). The unique linear sequence of amino acids that make up the polypeptide chain is known as the primary structure. Secondary structure of protein is formed by folding the pattern of polypeptides chains stabilized by hydrogen bonds. Hydrogen bonds are formed between amide hydrogens and carbonyl oxygen of the peptide backbone. Alpha (α) helix and beta (β) helix are the common types of secondary protein structures. The tertiary structure defines the spatial arrangement of the secondary structure. In a tertiary structure, interaction between the side chain functional groups, disulfide bonds, hydrogen bonds, salt bridges, and hydrophobic interactions stabilize the protein. This structure reflects the shape of the molecules and mostly consists of many smaller folds known as domain. This structure is determined by X-ray crystallography and nuclear magnetic resonance spectroscopy.

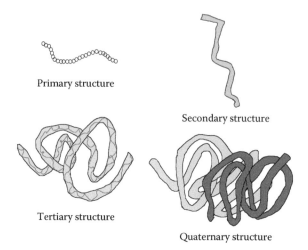

FIGURE 2.4
Protein structure at different levels of organization. (From Shamey, R. and Sawatwarakul, W., *Text. Prog.*, 46, 323–450, 2014. With Permission.)

The fourth level of protein structure consists of more than one polypeptide chain, and each chain structure is known as a subunit. The quaternary structure of the protein is related to the arrangement and interactions of the subunits. The dimer is the simplest kind of quaternary structure consisting of two identical subunits present in DNA-binding protein Cro in λ-bacteriophages. Complex quaternary-structured protein with more than one type of subunit, that is, α and β subunits is found in hemoglobin, the coat of rhinovirus is another example of complex quaternary structures (Berg et al. 2002).

2.6 Amino Acids

Amino acids are building blocks of proteins. In all organisms (bacteria, Archaea, and eukaryotes), different types of proteins are mostly made from the different combinations of 20 different amino acids. These 20 amino acids are known as a set of common or standard amino acids. Amino acids are amino derivatives of carboxylic acids in which both amino and carboxyl group are attached to α-carbon atom. All standard amino acids are therefore α-amino acids. A hydrogen atom and distinctive side chain for each amino acid are also attached to α-carbon atom. On the basis of the side chain, amino acid has specific characteristics, which determine its role in protein structure. Amino acids with aliphatic side chain include glycine, alanine, valine, leucine, and isoleucine; amino acids with aromatic side chain include

TABLE 2.1

Classification of Amino Acids Based on Their Properties

Property	Amino Acids
Hydrophobic	Alanine, valine, leucine, isoleucine, proline, phenylalanine, tryptophan, cysteine, and methionine
Positively charged	Lysine, arginine
Negatively charged	Aspartate, glutamate
Polar	Serine, threonine, asparagine, glutamine, histidine, and tyrosine
Small	Glycine, alanine
pK_a near neutrality	Histidine
Aromatic	Phenylalanine, tyrosine, and tryptophan
Hydroxyl side chain	Serine, threonine, and tyrosine
Helix bend or break	Proline

Source: Adapted from Schleif, R.F., *Genetics and Molecular Biology*, Johns Hopkins University Press, Baltimore, MA, 1993.

phenylalanine, tyrosine, and tryptophan; amino acids with sulfur containing side chains include methionine and cysteine; and alcohol containing side chain includes serine and threonine; positively charged side group containing amino acid includes histidine, lysine, and arginine; and negatively charged side group amino acid includes aspartate, glutamate, and their amide derivatives. On the basis of their properties, amino acids can be classified into many categories as shown in Table 2.1.

Protein consists of α-L-amino acids linked by peptide bonds to form a polypeptide chain. At neutral pH, the amino group is positively charged, whereas the carboxyl group is negatively charged. Thus, *N*-terminal of protein remains positively charged and the *C*-terminal is negatively charged. Positively and negatively charged amino acids often form salt bridges, which may be important for the stabilization of the protein 3D structure; for example, proteins from thermophilic organisms often have an extensive network of salt bridges on their surface, which contributes to the thermos ability of these proteins. Inside the cell, under normal physiological conditions at a pH range of 6.8–7.4, amino group ($-NH_3^+$) and carboxyl group ($-COO^-$) are ionized as they have pK_a (negative base-10 logarithm of the acid dissociation constant) value around 9 and 3, respectively. The amino acids thus exist in dipolar ion condition with 0 net charges, which is known as zwitterions. Except glycine, in rest of amino acids, the α-carbon atom is asymmetric or chiral, because four different groups are bonded to it. These 19 chiral amino acids exist as stereoisomers (same molecular formula but different arrangement) and enantiomers (nonsuperimposable mirror image of stereoisomers) (Moran et al. 2012).

2.6.1 The Peptide Bond

Amino acids are linked together in a polypeptide chain by peptide bond. The linkage between amino acids occurs through simple condensation reaction between the α-amino groups of one amino acid with α-carboxyl group of another amino acid with the release of a water molecule. Linked amino acids in a polypeptide chain are called amino acid residues. The free amino group and carboxyl group at the opposite ends of peptide chain are called the *N*-terminal and *C*-terminal, respectively. During protein synthesis, polypeptide chain formation starts from *N*-terminal of amino acid (usually methionine) and continues toward the *C*-terminals by adding one amino acid at a time. Depending on the number of amino acids linked together, they are termed as dipeptide, tripeptide, oligopeptide, and polypeptide. Dipeptide contains two amino acids linked by one peptide bond. Therefore, each peptide chain has one free amino and carboxyl group at opposite ends (Moran et al. 2012).

2.7 Polysaccharides and Lipids

Glucose, the simple sugar serves as major nutrient for cells, which is further metabolized to be utilized as cellular energy and as substrate for the synthesis of other cell constituents. Polysaccharides are storage forms of sugars: glycogen in animal and starch in plant, and form structural components of the cells. They act as marker for a variety of cell-recognition process, including the adhesion of cells and transport of proteins to appropriate intracellular destinations (Cooper 2000). Carbohydrates are one of the major biomolecules along with proteins, nucleic acids, and lipids. Carbohydrates are polymers of monosaccharides, which are compounds with multiple hydroxyl groups along with aldehyde or ketone. Monosaccharides are linked together in varying chain length and structure resulting in disaccharides, oligosaccharides, and polysaccharides. Carbohydrates in cells serve multiple functions, including energy source, structural unit of genetic material (DNA and RNA), and as building units of the cell structure such as cellulose in plant cell wall, one of the abundant material on the Earth. Carbohydrates can link together with proteins and lipids and assist in various cellular activities (Berg et al. 2002).

Fatty acids, long hydrocarbon chains (16–18C) with carboxyl group, are the simplest lipids and can be saturated or unsaturated depending on the absence or presence of unsaturated bonds between carbon atoms. Fatty acids with nonpolar C–H bonds exhibit hydrophobic nature, which is responsible for complex nature of lipids during cell membrane formation. Lipids, as one of the major biomolecules, serve three main functions in cells: first as energy storage, second as an important component of cell membrane, and finally in cell signaling, that is, steroid hormones (estrogen and testosterone) and messenger molecules conveying signals (Cooper 2000).

The membrane lipids are amphipathic: hydrophobic and hydrophilic moieties are present at opposite end of the membrane. Hydrophilic and hydrophobic interactions lead to packing of structural lipids into sheets called membrane bilayers. The major classes of structural lipids include the following: phospholipids composed of two fatty acids joined to polar head group along with characteristic phosphate group; glycerophospholipids composed of two fatty acids joined to glycerol and phosphate group bound to glycerol, galactolipids, and sulfolipids, which contain two fatty acids esterified to glycerol; tetraether lipids composed of two long alkyl chains linked to glycerol at both ends; and sphingolipids composed of fatty acids joined to fatty amine, sphingosine, and sterols (Figure 2.5) (Nelson and Cox 2012).

Phospholipids are the principal components of cell membranes. All phospholipids have hydrophobic tail consisting of two hydrocarbon chain and hydrophilic head comprising a phosphate group. In addition to phospholipids, cell membranes contain glycolipids and cholesterols. Glycolipids are composed of two hydrocarbon chains linked to polar head groups that contain carbohydrates. According to fluid mosaic theory (Figure 2.4), the plasma membrane is surrounded by two layers (a bilayer) of phospholipids. All phospholipid hydrophilic head is in contact with aqueous fluid both inside

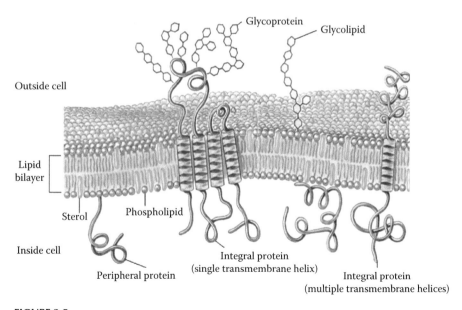

FIGURE 2.5
Fluid mosaic model. (From Nelson, D.L. and Cox, M.M., *Lehninger Principles of Biochemistry*, W.H. Freeman, New York, 2012. With Permission.)

and outside the cell, whereas hydrophobic tails are located inside the bilayer. Proteins are embedded in a bilayer by hydrophobic interaction between the membrane lipids. Some proteins protrude from only one side, whereas some protrude from both sides of the membrane leading to asymmetric orientation of protein in a bilayer. The individual structural lipid and protein in bilayer form fluid mosaic-like pattern (Cooper 2000).

References

Alberts, B., A. Johnson, J. Lewis, M. Raff, K. Roberts, and P. Walter. 2002. *Molecular Biology of the Cell*. New York: Garland Science.

Berg, J. M., J. L. Tymoczko, and L. Stryer. 2002. *Biochemistry*. New York: W. H. Freeman.

Cooper, G. M. 2000. *The Cell: A Molecular Approach*. Sunderland, MA: Sinauer Associates.

De La Rica, R. and H. Matsui. 2010. Applications of peptide and protein-based materials in bionanotechnology. *Chemical Society Reviews* 39:3499–3509.

Edberg, S. C. 1985. Principles of nucleic acid hybridization and comparison with monoclonal antibody technology for the diagnosis of infectious diseases. *The Yale Journal of Biology and Medicine* 58:425–442.

Griffiths, A. J. F., J. H. Miller, D. T. Suzuki, R. C. Lewontin, and W. M. Gelbart. 2000. *An Introduction to Genetic Analysis*. New York: W. H. Freeman.

Guo, P. 2010. The emerging field of RNA nanotechnology. *Nature Nanotechnology* 5:833–842.

Johnsburg, O., V. Eldholm, and L. S. Håvarstein. 2007. Natural genetic transformation: Prevalence, mechanisms and function. *Research in Microbiology* 158:767–778.

Levsky, J. M. and R. H. Singer. 2003. Fluorescence in situ hybridization: Past, present and future. *Journal of Cell Science* 116:2833–2838.

Liu, C. and A. D. Ellington. 2014. DNA nanotechnology: From biology and beyond BT. In *Nucleic Acid Nanotechnology* (Eds.) J. Kjems, E. Ferapontova, and K. V. Gothelf, pp. 135–169. Heidelberg, Germany: Springer.

Lodish, H., A. Berk, S. L. Zipursky, P. Matsudaira, P. D. Baltimore, and J. Darnell. 2000. *Molecular Cell Biology*. New York: W. H. Freeman.

Lorenz, M. G. and W. Wackernagel. 1994. Bacterial gene transfer by natural genetic transformation in the environment. *Microbiological Reviews* 58:563–602.

Meier-Kolthoff, J. P., A. F. Auch, H. P. Klenk, and M. Göker. 2013. Genome sequence-based species delimitation with confidence intervals and improved distance functions. *BMC Bioinformatics* 14:60.

Moran, L. A., R. A. Horton, K. G. Scrimgeour, and M. D. Perry. 2012. *Principles of Biochemistry*. Upper Saddle River, NJ: Pearson.

Nelson, D. L. and M. M. Cox. 2012. *Lehninger Principles of Biochemistry*. New York: W. H. Freeman.

Schleif, R. F. 1993. *Genetics and Molecular Biology*. Baltimore, MA: Johns Hopkins University Press.

Seeman, N. C. 2007. An overview of structural DNA nanotechnology. *Molecular Biotechnology* 37:246–257.

Shamey, R. and W. Sawatwarakul. 2014. Innovative critical solutions in the dyeing of protein textile materials. *Textile Progress* 46:323–450.

Swayze, E. E., R. H. Griffey, and C. F. Bennett. 2007. Nucleic acids (deoxyribonucleic acid and ribonucleic acid). In *Comprehensive Medicinal Chemistry II* (Eds.) J. B. Taylor and D. J. Triggle, pp. 1037–1052. Amsterdam, Netherlands: Elsevier.

Ullmann, F. 2007. Nucleic acids genetic engineering is a separate keyword. In *Ullmann's Biotechnology and Biochemical Engineering* (Ed.) Wiley-VCH Staff, pp. 1657–2187. Weinheim, Germany: Wiley-VCH.

Watson, J. D. and F. H. Crick. 1953. Molecular structure of nucleic acids. Nature, 171:737–38.

Wong, S. S. and D. M. Jameson. 2011. *Chemistry of Protein and Nucleic Acid Cross linking and Conjugation*. New York: CRC Press.

3

Genomics and Bionanotechnology

3.1 DNA Nanotechnology and Bionanotechnology

Nanotechnology is the emerging field, which makes use of both the metabolites and nano-objects synthesized by living beings. Different multidisciplinary technologies, which integrate different approaches, for instance, biology together with *omics*, has been successfully utilized for the development of nanostructured materials. Interaction between biological system and nanoparticles such as high diffusion rate and efficient uptake by living system, high surface-volume ratio, high biological impact due to mechanotransduction signaling and spectrum of alternative cell activities are the regulating factors of nanotechnology. Bionanotechnology, which is the integration of biology and nanotechnology, is usually observed as *nanotechnology through biotechnology* and involves biofabrication of nanoparticles (Villaverde 2010).

Molecular biotechnology is one of the emerging disciplines in nanotechnology and nanobiotechnology with the great potential to design well-defined structure nanomaterials utilizing DNA-guided materials. The exploration of the unique programmability, molecular recognition ability, and ability to undergo strand crossovers to make artificial objects of DNA molecules dates back to 1980s, when the crystallographer Seeman emerged as the founding father of DNA nanotechnology. He pioneered the idea of utilization of DNA molecules for the development of two-dimensional (2D) and three-dimensional (3D) structures at nanoscale (Schlichthaerle et al. 2016).

Nanotechnology based on DNA aims to develop user-defined objects in nanometer scales with high structural and functional complexities with the information encoded in DNA sequences. The most widely used approach for DNA nanotechnology fabrication involves connection of customized multiple double-helical DNA domains using strand backbone linkages. For instance, scaffolded DNA origami is the result of successful approach for

TABLE 3.1

Examples of DNA-Based Nanodevices with Their Application

DNA-Based Devices	Application
DNA nanotubes	For protein structure determination
DNA picture frames	For visualizing the conformation switching of G-quadruplex
2D DNA crystals	For visualizing single proteins
DNA origami gatekeepers	For sensing single-molecule stochastic
DNA chassis	For studying movements of molecular motor
DNA-based supports	For fabricating devices such as nanolens and polarizers
DNA-based fluorescent barcode	For identifying cells
DNA-based nanopill	For target delivery of molecular payloads to cells

Source: Adapted from Linko, V. and Dietz, H., *Cur. Opin. Biotechnol.*, 24, 555–561, 2013.

making objects containing thousands of DNA base pairs. During DNA origami designing, long single strand of DNA (weft yarn) is connected with specific sets of short DNA single strand (warp thread) to construct nano-objects with predetermined properties. Some of the nanodevices based on DNA are listed in Table 3.1.

3.2 Molecular Genetic Techniques Employing Nanotechnology

Molecular genetic techniques in combination with nanotechnology have revolutionized all fields of biological research. Recent advances in nanomaterials and nanofabrication technology have led to improvement in throughput, resolution, and information of the conventional genetic techniques. For instance, optical DNA mapping integrated with fluorescent labeling permits high-level genetic and epigenetic information on individual DNA molecules. Nanochannels introduced to extend DNA have the potential to improve the result of genome sequencing. DNA sequencing with nanopore technology has the potential to stochastically sense biomolecules with high throughput at permissive costs and speeds. Recently, the nanopore technology has been designed for application in genetic techniques such as DNA mapping, DNA sequencing, structural analysis, and protein detection (Jewett and Patolsky 2013).

Different DNA-conjugated nanoparticles such as DNA–gold nanoparticles (DNA–AuNP) conjugates and DNA–silver nanoparticles (DNA–AgNP) conjugates are important building blocks of DNA-based nanofabrication.

DNA nanotechnology in combination with gel electrophoresis techniques facilitates nanoseparation of DNA-conjugated nanoparticles due to the integral nanoporosity of agarose gel. Further, the isolated DNA-assembled nanomaterials have been identified to be be employed in agarose gel electrophoresis (Wang and Deng 2015).

3.3 DNA Amplification

DNA amplification is the process to copy number of sequences of DNA. Polymerase chain reaction (PCR) invented by Kary B. Mullis in 1980s is the first DNA amplification technique designed initially to study DNA in small quantities (Mullis 1990). With the advancements in PCR, it is possible to isolate and detect target DNA sequence, mutation or polymorphisms, paternity of child, genetic diseases, study human genome, and match DNA in forensic science. Different nucleic acid amplification techniques based on PCR have been modified for applications in molecular medicine. For instance, ligase chain reaction (LCR) is used for gene mutation characterization and detection of infectious diseases; nucleic acid sequence-based amplification (NASBA) and related transcription-mediated amplification are used for infection diagnosis such as human immunodeficiency virus (HIV), hepatitis, and branched DNA-based detection for diagnostic purpose (Sorscher 1997).

Due to its simplicity, easier methodology, and availability of chemicals and reagents, and valid standard operating procedure, PCR is the most preferred method for nucleic acid amplification. Limitations associated with PCR method are expensive equipment and operating costs, chances of contamination, requirements of thermal cycling, and sensitivity to inhibitors. The alternative molecular methods such as LCR, NASBA, rolling circle amplification (RCA), loop-mediated isothermal amplification (LAMP), self-sustained sequence replication (3SR), and strand displacement amplification (SDA) have also been developed to analyze DNA, RNA, and other similar molecules (Fakruddin et al. 2013).

PCR is the classical technique used to amplify DNA *in vitro* by mimicking the cellular process of DNA replication. This method is used for isolating specific DNA, DNA end labeling, cDNA and genomic DNA cloning, DNA sequencing, RNA and DNA quantification, study of gene expression, and to mutagenize target DNA sequence with wide applications in forensic and diagnostic fields. Molecular biotechnology has undergone vast improvement and development revolutionizing basic biology, systematics, ecology, and evolution leading to the applications at molecular levels to diverse problems. The method such as PCR involves complex kinetic interactions between the template DNA (target DNA), product DNA, oligonucleotide

TABLE 3.2

Reagents and Equipments Required for the Polymerase Chain Reaction

Requirement	Specification
Target or template DNA	The specific segment of nucleic acid to be amplified
Primer	Short single-stranded polymer of oligonucleotide that anneals to target DNA by complementary base pairing
Taq DNA polymerase	Heat-stable enzymes that make new complimentary copy of target nucleic acid by adding nucleotides to annealed primer
Reverse transcriptase	Enzyme to convert RNA into a complementary DNA sequence
Thermocycler	Equipment used in PCR reaction that can change rapidly to the different temperatures required for repeated PCR cycles

Source: Adapted from Louie, M. et al., *Can. Med. Assoc.*, 163, 301–309, 2000.

primers (10–30 nucleotide polymer), deoxyribonucleotide triphosphates (dNTP), buffer ($MgCl_2$), and enzyme (one or more DNA polymerases) (Table 3.2). The optimum PCR process depends on the process parameters, that is, reaction buffer, concentration of target DNA, primers, dNTPs and DNA polymerase, annealing time and temperature, and extension time and temperature (Hoy 2013).

Nucleic acid amplification is done in three steps, that is, denaturation, annealing, and extension (Figure 3.1). In first step, target DNA is heated to around 90°C–95°C causing denaturation of DNA due to which double-stranded DNA separates into single strands. After the DNA has been denatured, in annealing step the separated strands are cooled to 55°C and the oligonucleotide primers hybridize to the denatured DNA at the specific targets. During extension step, *Taq* DNA polymerase sequentially adds

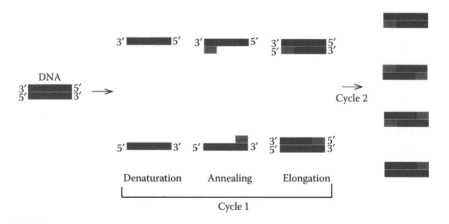

FIGURE 3.1
Schematic representation of the polymerase chain. (From Garibyan, L. and Avashia, N., *J. Invest. Dermato.*, 133, 1–4, 2013. With Permission.)

the dNTPs to the template DNA strands, synthesizing two new identical double-stranded DNA molecules. These three steps together complete one cycle and the process is repeated for several cycles so that at the end of each cycle, newly synthesized DNA sequence acts as a new target for the next cycle. After completion of 30 cycles, millions of copies of original target DNA are created resulting in accumulation of specific PCR products (Louie et al. 2000).

The products can further be analyzed by two methods: (1) staining with chemical dye, ethidium bromide that intercalates between the two strands of the duplex and (2) labeling the PCR primers or nucleotides with fluorescent dyes, fluorophores prior to PCR amplification that cause direct addition of labels in the PCR products. Agarose gel electrophoresis is most widely used for analyzing the PCR products, in which DNA products are separated based on size and charge. The size of the product is determined by simultaneously running a predetermined set of DNA products as molecular markers. PCR approach can be either qualitative that detects the presence or absence of specific DNA or quantitative that indicates about quantity of specific DNA. These types of techniques of nucleic acid amplification have the advantages of being simple, rapid, and sensitive method with the potential to produce millions to billions copies of specific produce for sequencing, cloning, and analyzing. It is also useful to analyze alteration of gene expression levels in tumors, microbes, or diseased conditions. On the other hand, this method presents few limitations that are listed as follows (Garibyan and Avashia 2013):

1. Need of special machine with automated thermal cycle.
2. High processing temperature may damage original chromosome structures.
3. Due to high sensitivity of the process, results might be misleading in the case of minute contamination.
4. Need of primers implies that only known genes can be identified.
5. In some cases, primers may anneal nonspecifically to similar sequences leading to mismatched base pairing.

3.3.1 Application of Polymerase Chain Reaction

3.3.1.1 Reverse Transcription-Polymerase Chain Reaction

PCR can be applied for the quantification of low level of gene expression by utilizing RNA as template. Reverse transcription-PCR (RT-PCR) is the modification of PCR where single-stranded RNA is used as initial template, and enzyme (reverse transcriptase) converts target RNA into complementary DNA (cDNA) copy, which is amplified further by standard PCR method. This approach can be used to amplify messenger RNA (mRNA) or ribosomal RNA (rRNA) rather than DNA, thus detecting specific expression of certain genes during the state of infection. RT-PCR method is therefore useful to detect infection by detecting cDNA of mRNA encoded by a pathogen (Louie et al. 2000).

3.3.1.2 In Situ *PCR*

Histological tissue samples can be analyzed *in situ* by PCR analysis for the detection of infectious organisms, such as HIV. Ideally, it is performed in deep well slide where tissue section/cytology smear is sealed with nail varnish, rubber, or agarose for complete compartmentalization of the PCR reaction within the slide. The PCR products can be detected by colorimetric methods or by microscopic inspection, which can identify particular target sequence (O'Leary 1994).

3.3.1.3 PCR in Molecular Genetics

PCR is a versatile approach used to resolve diagnostic, ecological, evolutionary, genetic, and developmental biology problems. PCR can be utilized to amplify ancient DNA such as DNA fragments of insects preserved in ancient amber, dinosaur DNA, mitochondrial DNA, mammoth DNA, and DNA from museum specimens such as Lyme disease pathogen DNA. PCR is useful to amplify mRNA to study gene expression during development, quantitation of mRNA, rearrangements of DNA during cell differentiation and RNA processing, and for analysis of mRNA polyadenylation to study mRNA stability. Other applications of PCR include DNA sequencing, DNA engineering, detection of amplified gene, detection of methylation of DNA, detection of infectious pathogens, detection of pesticide resistance, evaluation of efficacy of disease-control mechanism, developmental biology, DNA engineering, and analysis of evolution (Hoy 2013). In addition, PCR is also utilized for gene cloning or gene fragmentation into plasmid vectors, which can be subsequently used for protein expression or overexpression and subsequent purification of proteins (Maddocks and Jenkins 2017).

3.4 Gene Cloning

Gene cloning and vector construction are widely applied techniques in DNA and protein research and molecular biology. Gene cloning is defined as the process of isolating gene or DNA of interest using restriction enzyme or PCR method, ligating the isolated gene into suitable vector and production

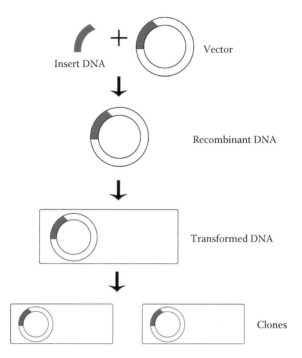

FIGURE 3.2
General scheme of gene cloning. (From Wong, D.W.S., *The ABCs of Gene Cloning,* Springer US, Boston, MA, 2006. With Permission.)

of enough copies of exact vectors. Usually, plasmids are used as vectors that are usually small circular DNA molecules physically different from chromosomal DNA and capable of independent replication. It acts as vehicle for introducing the gene into host cell and directs the gene replication and expression. The vector DNA with foreign DNA is known as recombinant DNA and the process of introducing foreign gene-containing vector into host cell is called transformation (Figure 3.2). The transformed cell or the host cell with gene-containing vector produces identical cells known as clones (Wong 2006).

3.4.1 Steps of Gene Cloning

Broadly gene cloning can be done in four basic steps: (1) DNA sequence of interest is fragmented into smaller pieces to prepare insert DNA, (2) insert DNA fragment is joined to vector DNA, (3) recombinant DNA molecule is inserted into a host cell where it gets replicated and multiple copies of cloned

DNA are produced, and (4) cells containing recombinant clones are identified and further propagated (Stephenson 2010).

The general procedure of gene cloning involves *in vitro* assembly of recombinant DNA followed by insertion to host cell, which can direct the replication of recombinant DNA in coordination with its growth. Nonpathogenic bacterial strain of *Escherichia coli* is used as host cell, which can grow exponentially to generate virtually unlimited identical copies of the target DNA. Vectors are the DNA fragments with origin of plasmid replication and the DNA fragment with no origin of replication, which is being joined to vector is known as insert. Insertion into a cloning vector and the exchange of a vector DNA fragment for an insert DNA fragment are the standard DNA cloning approaches.

DNA cloning can be either single step or series of steps united with cloning strategy. The multistep approach includes (Tolmachov 2011):

1. Insertion of multiple cloning site also known as polylinker to facilitate subsequent cloning
2. Subcloning of the insert inside multiple cloning site to change restriction sites
3. Subcloning to change antibiotic selection marker on the recombinant plasmid
4. Stepwise insertion of individual genetic elements
5. Two-step pop-in-pop-out strategy where insert is transferred from one to another vector

3.4.2 Methods of Gene Cloning

The conventional method of gene cloning starts with the action of restriction endonucleases, an enzyme that cleaves DNA and plasmid at specific sequence. The cleaved DNA fragments may be either with *sticky ends* in which the DNA fragment has single-stranded overhang (either on 3′ or 5′ end) or *blunt ends* with no overhang. The cleaved target insert sequence and plasmid are stitched together by the enzyme DNA ligase.

The classical technique is convenient in the presence of well-defined restriction site in the sequence. On the other hand, this approach is not suitable in case of absence of restriction site and most of the eukaryotic genes are interrupted by intervening sequences (introns), which makes the gene of interest very large. As manipulation of the large genomic DNA is difficult due to size capacity of cloning vectors and multiple restriction endonucleases, alternative cloning approaches such as PCR-mediated TA cloning, ligation-independent cloning (LIC), recombinase-dependent cloning, and PCR-mediated cloning are being developed (Souii et al. 2013).

- In PCR-mediated TA cloning method, DNA fragment generated from PCR is directly stitched with vector without the use of any restriction enzymes. As DNA fragments with adenine (A) residue known as *A-tailed* DNA fragment and vectors with thymine (T) residue known as *T-tailed* are directly ligated, this approach is known as *TA* cloning.
- LIC is done by adding short sequence of DNA to the insert DNA, which is homologous to the vector. Complementary cohesive ends between the insert and vector are formed by the enzymes with 3′–5′ exonuclease activity and the resulting two molecules are annealed together such that plasmid will have four single-stranded DNA nicks.
- The seamless cloning method allows sequence-independent and scar-free, that is, free of restriction enzyme sites and unwanted sequence, insertion of DNA insert into vector. This method includes addition of homologous region at each end of insert to be cloned and with action of enzyme exonuclease, DNA polymerase and ligase-recombinant DNA are formed.
- Recombinatorial cloning technique utilizes site-specific DNA recombinase enzymes that can swap pieces of DNA between two molecules with appropriate recombination site. Initially, the recombination sites are added on either side of insert by PCR followed by recombination with vector to make an entry clone, which is further recombined with destination vector to develop final construct. Gateway cloning system is one of the widely used cloning techniques based on this approach (Bertero et al. 2017).

3.4.3 Application of Gene Cloning

Gene cloning is an important tool to determine the DNA sequence and function of both coding genes and noncoding elements of the genome. Cloning of complementary DNA (cDNA) into an expression vector to induce overexpression in a target organism or cloning of specific short-hairpin RNA (shRNA, short sequence that can suppress gene expression) is applicable to study the function of specific gene sequence. Specific mutation induced by site-directed mutagenesis is used to assess gene functions based on molecular cloning strategy. *In vitro* or *in vivo* cloning of putative gene promoter or enhancers into specific vector is done to study the function of noncoding elements of genes. Gene cloning is an important tool with wide biomedical translational applications such as development of recombinant proteins for diagnostic or therapeutic purpose. The transgenic plants and animals are developed with the aim of introducing gene of interest (cloning) for pharmaceuticals' nutritional or biological value (Bertero et al. 2017).

3.5 Recombination of DNA

DNA recombination strategy is the technique to produce DNA via artificial means. The natural DNA sequence of the living organisms, microbes, plants, and animals is changed by recombinant DNA technology with the aim of expressing new characteristics. Thus the DNA resulting from the combination of the two DNAs of different origins is known as recombinant DNA (Stryjewska et al. 2013).

Recombinant DNA (rDNA) molecules are formed by molecular cloning (genetic recombination), which create a new sequence in the DNA of biological organisms by combining genetic material from multiple sources. It is possible to stitch the genetic material of different organisms because DNA molecules from all sources share same chemical structures, that is, all DNAs are made up of base consisting of pentose sugar, phosphate, and one of four nitrogen bases but differ in the nucleotide sequence, which can be arranged in infinite ways in the double helix structure (Zimdahl 2015).

The goals of application of recombinant DNA technology vary with the field of application, which are summarized in Table 3.3.

TABLE 3.3

Summary of Application of Recombinant DNA Technology in Different Fields

Purpose	Recombinant Product	Reference
Agriculture		
To slow ripening, slow softening, and preserve color and flavor	Flavr Savr tomato	Zimdahl (2015)
To produce virus-resistant plants	Papaya, potato, plum, and squash	Eck (2013)
To improve nutrient profiles	High-lysine corn and golden rice	
Food technology		
Microbial production of enzyme for cheese production	Chymosin	Zimdahl (2015)
Production of enzymes and proteins	Acetolactate decarboxylase, laccase, amylase, aminopeptidase, lipase, pullulanase, phytases, pectinase, asparaginase, carboxypeptidase, cellulase, glucose oxidase, hexose oxidase, and ice-structuring protein	Eck (2013)
Pharmaceutical application		
For diabetic	Insulin	Stryjewska et al. (2013)
For growth hormone deficiency	Human growth hormone	
For anemia treatment	Erythropoietin	
For prevention of blood clotting	Activase	
For the development of live-attenuated vaccines	Tetravalent dengue (DEN) vaccine	Lee et al. (2012)

3.6 DNA Sequencing

DNA molecule is composed of deoxyribose sugar, phosphate group, and four different types of nitrogenous bases, namely, cytosine, adenine, guanine, and thymine. DNA sequencing, therefore, is the study of the arrangement of these nucleotides in a strand of DNA molecules, which regulates the blueprint of life and provides genetic and biological information at the molecular level. DNA sequencing involves sequencing of genomic DNA, complementary DNA (cDNA), and analysis of epigenetic modification of genomic DNA molecules (Mitsui et al. 2015). With the advancement in molecular biology, DNA sequencing technique also experiences improvement, and the DNA sequencing methods are classified into four generations.

Sanger's chain termination or dideoxy first generation technique involves utilization of DNA polymerase, single-stranded DNA template, primer DNA, chemical analogs of the deoxyribonucleotides (dNTPs), and dideoxynucleotides (ddNTPs). The molecule ddNTPs lack 3'hydroxyl, which is needed to bond with 5' phosphate of the next dNTP molecule, thus leading to the termination of extension. Briefly, the process starts with annealing of primer to specific region of template DNA, which starts to synthesize DNA strand in the presence of DNA polymerase. During DNA extension reaction, addition of radiolabeled ddNTPs at fraction of the concentration of standard dNTPs results in random incorporation of ddNTPs in DNA strand terminating the extension. The experiment is conducted in four tubes each with specific amount of the four dNTPs and the resulting fragments with same 5'-end, whereas 3'-end is determined by specific ddNTP, which is run on four lanes of polyacrylamide gel. Nucleotide sequence in the original template is determined by autoradiography, which produces a radioactive band in the corresponding lane (Heather and Chain 2016).

The improvement of large-scale dideoxy sequencing efforts sets the stage for the development of second-generation DNA sequencing technique, which relied on luminescent method for measuring pyrophosphate synthesis, enzyme that catalyzes conversion of pyrophosphate into ATP. Amplification-based parallel sequencing method involves *in vitro* amplification of DNA strands and their clustering onto surfaces as well as the sequencing via synthesizing the arrays of microbeads, DNA, nanoballs, and DNA clusters. Third-generation single molecule sequencing involves stepwise and real-time single-molecule sequencing by synthesis, single-molecule motion sequencing, Raman scattering-based sequencing, electron microscopy, molecular force spectrometry, and others. Nanopore-based fourth-generation DNA sequencing is expected to relate genomic science into personal medicine. This technique is more reliable, cheap, and provides high-throughput sequencing. Companies such as Oxford Nanopore Technologies, Life and Roche with IBM are involved in research and development of nanopore sequencing (Yang and Jiang 2017).

Novel sequencing techniques such as *in situ* sequencing that involves *in situ* sequencing of mRNA by ligation; transmission electron microscopy (TEM) sequencing that involves labeling of DNA bases by tags with different contrasts; and electronic sequencing that involves immobilization of nucleic acid-coated beads, reagents, and sequencing beads by the application of magnetic and electronic fields are at different stages of development (Kulkarni and Pfeifer 2015). DNA sequencing by different techniques such as tunneling current, mechanical identification, mass spectrometry, electron microscopy, RNA polymerase sequencing, fluorescence resonance energy transfer-based and Raman spectroscopy approaches are the future of sequencing techniques (Ansorge et al. 2017).

3.6.1 Application of Gene Sequencing

Nucleic acid sequencing is routinely used to detect and characterize virus, to investigate host–pathogen relation, and to develop effective disease treatment approaches. Metagenomic approach used for identification and sequencing of target virus is based on analyzing infected host tissue as it is difficult to isolate and culture the pathogens. This method utilizes transcriptomic response of host and virus to detect their possible interaction and to develop novel drugs for treatment and drug targets (van Aerle and Santos 2017).

Next-generation sequencing technologies have revolutionized high-throughput functional genomic research with wide applications in the whole-genome sequencing, targeted sequencing, genome annotation, identification of transcription factor binding sites, identification of DNA sequences associated with epigenetic modification of histones and DNA, profiling of noncoding RNA expression, gene expression, posttranslational modification of histones, DNA methylation, and nucleosome positions on genome (Morozova and Marra 2008).

Study of epidemiology and public health is one of the major areas of application of whole genome sequencing where it is important to detect reasons of outbreak, to monitor the evolution dynamics of multidrug-resistant and novel antibiotic-resistant pathogens, to characterize highly virulent bacteria, and to implement the control measures. Gene sequencing databases can further retrospectively be investigated in case of complex and comprehensive outbreaks. Gene sequencing serves as a suitable one-step protocol to characterize broad range of pathogens instead of conventional bacteriological, biochemical, and molecular methods. Early detection of virulence profile of a pathogen is important to predict the severity of disease and infection outcome permitting the risk assessment during the onset of the disease. Gene sequencing further allows to investigate transmission of zoonotic microorganisms from animals to humans (Deurenberg et al. 2017).

3.7 Genome Mapping

Genome mapping is the process of locating the order of gene and their relative distance on the genome. It provides guidelines for the reconstruction of genome sequence after sequencing. DNA sequencing is a complex process due to which the genome needs to be fragmented before sequencing. The fragmented genome sequences are rearranged in original order using recognizable features. As genome map carries information about genome organization in terms of genes, restriction enzyme sites, and others, genome map is utilized to reconstruct the original genome after sequencing. There are two types of genome mapping that includes genetic mapping and physical mapping. For genetic mapping, commonly used markers are genes with visible phenotype and molecular markers, which is DNA sequence that shows polymorphism. Physical mapping involves direct location of DNA sequence on the chromosome using genome-wide unique DNA sequences, sequence-tagged site (STS), and expressed sequence tag (EST) as markers. Different physical mapping techniques include cytogenetic mapping, fluorescent *in situ* hybridization (FISH), restriction mapping, STS content mapping, and radiation hybrid mapping (Saraswathy and Ramalingam 2011).

3.8 Human Genome Project

The term *genome* was coined in 1920 by a German geneticist Winkler to describe the complete haploid set of chromosomes. Since then, genome refers to the complete genetic information either DNA or RNA sequence of any living organism. In early 1980s, the concept of positional cloning was developed based on detection of the position of genes in the genetic map responsible for inherited genetic diseases. By 1986, there was sufficient development of technologies for DNA cloning and sequencing for possible sequencing of complete human genome. In 1987, the term *genomics* was used first by McKusick and Ruddle and defined as "the newly developing discipline of genome sequencing including analysis of the information." Initially, Human Genome Project (HGP) was developed to obtain complete sequence of the 3 billion base pairs of DNA along the chromosomes. Initially map-based cloning sequencing strategy using bacterial artificial chromosomes (BACs) with insert lengths of approximately 150,000 base pairs is used for sequencing. In 1998, Celera Genomics, a private venture led by Venter, introduced an alternative to BAC mapping phase known as *shotgun* approach that works with relatively short sequence (Bodmer 2013).

HGP was initiated in 1990 with the aim of determining the order of nucleotides in human genome, which is made of 23 pairs of chromosomes, each with euchromatic and heterochromatic regions. The major goals of HGP were to prepare high-resolution human genome map using genetic and physical mapping approaches, to determine arrangement of nucleotide in chromosomes, to develop high throughput sequencing methods, to know genome sequence of test organisms to test feasibility of available sequencing and mapping technologies, to develop automated sequencing techniques, to annotate DNA sequence based on sequence content, and finally to address consequences of genomic research through ethical, legal, and social implications (ELSI) program (Saraswathy and Ramalingam 2011).

In April 2003, HGP was officially announced to be completed with a high-quality sequence of the entire human genome. In addition to the human genome, the HGP sequenced the genomes of several other organisms, including brewer's yeast, roundworm, and fruit fly. By studying the similarities and differences between human genes and those of other organisms, researchers discovered the functions of particular genes and identified the genes that are critical for life (Falcón de Vargas 2002).

The HGP provides the basis for functional genomics, which implies about study of functional role of the genomes and interaction of genes and their products in biological function and in diseased conditions. Based on genomic arrangement and designated functional role, new therapeutic proteins are being developed. HGP also helped to characterize the genomes of different pathogens and others in order to improve human health. Analysis of human genome sequence showed that 5.3% of the euchromatic regions carry segmental duplication, which is responsible for deletion and rearrangements of chromosomal regions that result in many genetic diseases such as Williams syndrome, Charcot-Marie-Tooth region, and DiGeorge syndrome region. Response to drugs depends on genetic make-up, for example, same dosage of drug has different responses on patients with similar disease. Human genome resources can be utilized to optimize the drug dosage for each individual patient for personalized medicine and to identify genes involved in graft rejection during organ transplantation (Saraswathy and Ramalingam 2011).

The availability of a reference human genome sequence has wide range of applications in the identification of genes responsible for diseases such as Huntington's disease, chronic granulomatous disease, Duchenne muscular dystrophy, and retinoblastoma. Information from HGP has been used to identify genes of biomedical importance such as breast cancer susceptibility, total color blindness, and X-linked lymphoproliferative disease. Human genome has wide applications in genome-wide association (GWA) studies, which is defined as "any studies of common genetic variation across the entire human genome designed to identify genetic associations with observable traits." GWA aims to provide guidance for risk assessment, diagnosis, and drug therapies (Kumar and Upadhyaya 2014).

References

Ansorge, W. J., T. Katsila, and G. P. Patrinos. 2017. Perspectives for future DNA sequencing techniques and applications. In *Molecular Diagnostics* (Eds.) G. P. Patrinos, W. J. Ansorge, and P. B. Danielson, pp. 141–153. Tokyo, Japan: Academic Press.

Bertero, A., S. Brown, and L. Vallier. 2017. Methods of cloning. In *Basic Science Methods for Clinical Researchers* (Eds.) M. Jalai, F. Y. L. Saldanha, and M. Jalali, pp. 19–39. Tokyo, Japan: Academic Press.

Bhatia, S. 2015. Application of plant biotechnology. In *Modern Applications of Plant Biotechnology in Pharmaceutical Sciences* (Eds.) S. Bhatia, K. Sharma, R. Dahiya, and T. Bera, pp. 157–207. Tokyo, Japan: Academic Press.

Bodmer, W. 2013. Human genome project. In *Brenner's Encyclopedia of Genetics* (Ed.) M. Caplan, pp. 552–554. Tokyo, Japan: Academic Press.

Deurenberg, R. H., E. Bathoorn, M. A. Chlebowicz et al. 2017. Application of next generation sequencing in clinical microbiology and infection prevention. *Journal of Biotechnology* 243:16–24.

Eck, P. 2013. Recombinant DNA technologies in food. In *Biochemistry of Foods* (Eds.) N. A. M. Eskin and F. Shahidi, pp. 503–556. Tokyo, Japan: Academic Press.

Fakruddin, M., K. S. B. Mannan, A. Chowdhury et al. 2013. Nucleic acid amplification: Alternative methods of polymerase chain reaction. *Journal of Pharmacy and BioAllied Sciences* 5:245–252.

Falcón de Vargas, A. 2002. The human genome project and its importance in clinical medicine. *International Congress Serie*s 1237:3–13.

Garibyan, L. and N. Avashia. 2013. Polymerase chain reaction. *Journal of Investigative Dermatology* 133:1–4.

Heather, J. M. and B. Chain. 2016. The sequence of sequencers: The history of sequencing DNA. *Genomics* 107:1–8.

Hoy, M. A. 2013. *Insect Molecular Genetics*. Tokyo, Japan: Academic Press.

Jewett, M. C. and F. Patolsky. 2013. Nanobiotechnology: Synthetic biology meets materials science. *Current Opnion in Biotechnology* 24:551–554.

Kulkarni, S. and J. Pfeifer. 2015. *Clinical Genomics*. Tokyo, Japan: Academic Press.

Kumar, A. and K. C. Upadhyaya. 2014. Perspectives on the human genome. In *Animal Biotechnology* (Eds.) A. S. Verma and A. Singh, pp. 577–595. Tokyo, Japan: Academic Press.

Lee, H.-C., M. Butler, and S. C. Wu. 2012. Using recombinant DNA technology for the development of live-attenuated dengue vaccines. *Enzyme and Microbial Technology* 51:67–72.

Linko, V. and H. Dietz. 2013. The enabled state of DNA nanotechnology. *Current Opinion in Biotechnology* 24:555–561.

Louie, M., L. Louie, and E. Simor. 2000. The role of DNA amplification technology in the diagnosis of infectious diseases. *Canadian Medical Association* 163:301–309.

Maddocks, S. and R. Jenkins. 2017. *Understanding PCR*. Tokyo, Japan: Academic Press.

Mitsui, J., H. Ishiura, and S. Tsuji. 2015. DNA sequencing and other methods of exonic and genomic analyses. In *Rosenberg's Molecular and Genetic Basis of Neurological and Psychiatric Disease* (Eds.) R. N. Rosenberg and J. M. Pascual, pp. 77–85. Tokyo, Japan: Academic Press.

Morozova, O. and M. A. Marra. 2008. Applications of next-generation sequencing technologies in functional genomics. *Genomics* 92:255–264.

Mullis, B. 1990. The unusual origin of the polymerase chain reaction. *Scientific American* 262:56–61.

O'Leary, J. 1994. The polymerase chain reaction and fixed tissues. In *Methods in DNA Amplification* (Eds.) A. R. I. Weber-Rolfs and U. Finckh, pp. 3–9. Boston, MA: Springer US.

Saraswathy, N. and P. Ramalingam. 2011. *Concepts and Techniques in Genomics and Proteomics*. Oxford, UK: Woodhead Publishing.

Schlichthaerle, T., M. T. Strauss, F. Schueder, J. B. Woehrstein, and R. Jungmann. 2016. DNA nanotechnology and fluorescence applications. *Current Opinion in Biotechnology* 39:41–47.

Sorscher, D. H. 1997. DNA amplification techniques. In *Molecular Diagnostics: For the Clinical Laboratorian* (Eds.) W. B. Coleman and G. J. Tsongalis, pp. 89–101. Totowa, NJ: Humana Press.

Souii, A., M. M'Hadhed-Gharbi, and J. Gharbi. 2013. Gene cloning: A frequently used technology in a molecular biology laboratory—Alternative approaches, advantages. *American Journal of Research Communication* 1:18–35.

Stephenson, F. H. 2010. *Calculations for Molecular Biology and Biotechnology*. Tokyo, Japan: Academic Press.

Stryjewska, A., K. Kiepura, T. Librowski, and S. Lochyński. 2013. Biotechnology and genetic engineering in the new drug development. Part I. DNA technology and recombinant proteins. *Pharmacological Reports* 65:1075–1085.

Tolmachov, O. E. 2011. *Comprehensive Biotechnology*. Tokyo, Japan: Academic Press.

van Aerle, R. and E. M. Santos. 2017. Advances in the application of high-throughput sequencing in invertebrate virology. *Journal of Invertebrate Pathology* 147:145–156.

Villaverde, A. 2010. Nanotechnology, bionanotechnology and microbial cell factories. *Microbial Cell Factories* 9:53.

Wang, H.-Q. and Z. X. Deng. 2015. Gel electrophoresis as a nanoseparation tool serving DNA nanotechnology. *Chinese Chemical Letters* 26:1435–1438.

Wong, D. W. S. 2006. *The ABCs of Gene Cloning*. Boston, MA: Springer US.

Yang, N. and X. Jiang. 2017. Nanocarbons for DNA sequencing: A review. *Carbon* 115:293–311.

Zimdahl, R. L. 2015. *Six Chemicals That Changed Agriculture*. Tokyo, Japan: Academic Press.

4

Protein Engineering and Bionanotechnology

4.1 Protein Engineering and Bionanotechnology

Different types of proteins with specific structure and functions are widely available in nature. Protein engineering is defined as the technique of designing new enzyme or protein with specific function. This technique is based on DNA recombination technology to rearrange the sequence of amino acids and high-throughput screening technique to increase the enzyme stability (Turanli-yildiz et al. 2012). Protein engineering is getting importance for two reasons: (1) to gain more information about natural proteins, which is achieved by developing new proteins with the same functionality and (2) to develop novel macromolecules to solve problems in chemical and biochemical fields. Broadly, the working definition of protein engineering can be divided into four classes (Galzie 1991):

Class 1: It implies the experiment on the system that can make a successful application or has a chance to succeed by utilizing satisfactory relationship from primary sequence of information.

Class 2: The second class is based on the experiment related to screening of the mutated genes from random gene pool to obtain desired proteins.

Class 3: The third class is based on structure–function relationship.

Class 4: The fourth class is composed of experiments for improving protein characteristic by designing *de-novo* protein.

The definition of protein engineering has undergone modifications with the development achieved in different fields. Classically, protein engineering involves either rational design or directed evolution approaches to modify protein properties. With the increased knowledge of structural genomics

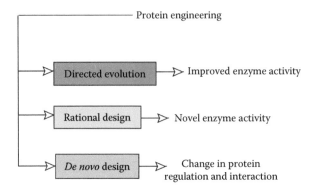

FIGURE 4.1
General approach of protein engineering. (From Chen, Z. and Zeng, A.P., *Curr. Opin. Biotechnol.*, 42, 198–205, 2016. With Permission.)

and computational approach such as protein sequences, structure, evolution, dynamics, and function, protein engineering functions in development of efficient biocatalytic process (Figure 4.1) (Chen and Zeng 2016).

4.2 Protein Engineering Methods

Many new methods for protein engineering are emerging owing to the development achieved in recombinant DNA technology. The protein engineering methods are classified into two major groups: (1) rational method and (2) random method. Rational method involves site-directed mutagenesis, whereas random method includes random mutagenesis and evolutionary methods. Rational method and directed evolution are the most common tools used for protein engineering (Turanli-yildiz et al. 2012).

4.2.1 Rational Design as Nanomaterials

The rational design of protein engineering for fabrication at nanoscale is the classical method that uses the biochemical principles and knowledge from past studies to create proteins with new structure and functionalities. An advanced form is computational protein design that uses computer modeling to design new proteins or enzymes and to foresee how amino acid chains would fold to form a protein (Tobin et al. 2014). In this technique, the general backbone of protein must be preserved. However, some flexibility can be added to the original backbone to increase the possible sequence that folds to the structure and maintaining the general fold at the same time (Mandell and Kortemme 2009).

The rational design uses site-directed mutagenesis where the amino acid is injected into a target gene. Two common methods for site-directed mutagenesis are: (1) overlapped extension method and (2) whole plasmid single-round polymerase chain reaction (PCR). In overlap extension method, two primer pairs and short sequences of synthetic DNA complementary to gene of interest are involved in the first PCR wherein two separate PCRs are performed. First PCR uses four primers resulting in two double-stranded DNAs, which on denaturation and annealing produce two heteroduplexes with desired mutagenic codon in each strand. The overlapping in 3' and 5' ends of each heteroduplex is filled by DNA polymerase and then the mutagenic DNA is amplified by second PCR using a nonmutated primer. After DNA polymerase PCR takes place, the desired mutated plasmid without overlap in it breaks due to replication of strands template. In the whole plasmid single-round PCR method, two oligonucleotide primers with desired sequences and complementary to the opposite strands of double-stranded DNA plasmid template are extended using DNA polymerase. During PCR, both strands are replicated without displacing the primers. Circular mutated plasmid is obtained upon transformation of the circular nicked vector (Antikainen and Martin 2005).

4.2.2 Directed Evolution of Protein Designing as Nanomaterials

Directed evolution is one of the most powerful protein engineering methods, which is used to redesign the protein structure to modify or create new protein with desired attribute, for instance, improving the stability or activity of enzyme under extreme condition through multiple mutations. In this method, random or focused mutagenesis and selection are proven to be effective even under the condition of limited information regarding the structure and mechanism of protein (Verma et al. 2012). Conventionally, directed evolution consists of two steps: (1) the first step is to produce molecular variety by random mutagenesis and DNA recombination and (2) the second step is applying high-throughput screening to identify the proper sequences for their functionalities (Figure 4.2). There is availability of protein library with collection of information of millions of proteins. Modern approaches based on available protein structure, sequence, and function together with computational predictive algorithms are being utilized for protein engineering (Lutz 2010).

Evolutionary methods are being used to change the protein properties including substrate selectivity. *In vitro* evolution method is used to create new protein with specific functions, involving randomization of the entire target gene by error-prone PCR followed by multiple screening and selection. *DNA shuffling* is the direct evolution method where genes with similar

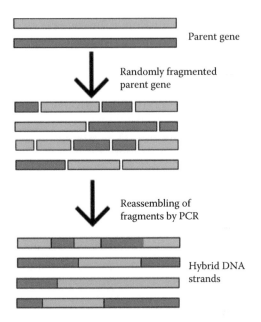

FIGURE 4.2
Schematic representation of DNA shuffling. (From Antikainen, N.M. and Martin, S.F., *Bioorgan. Med. Chem.*, 13, 2701–2716, 2005. With Permission.)

sequence double-stranded DNA are produced from organisms or by error-prone PCR. Genes are randomly digested into small fragments by *DNase* I, which are further purified and reassembled using an error-prone and thermostable DNA polymerase through PCR. During PCR, the fragments act as primers, which align and cross-prime each other resulting in hybrid DNA (Turanli-yildiz et al. 2012).

4.3 Protein Engineering Applications

Applications of protein engineering have been reported in different disciplines ranging from food industry, environmental science, medical biotechnology, and nanobiotechnology. Different protein engineering methods, rational design, directed evolution, and diversity of the nature alone or in combination have greater scope of the application in different fields as summarized in Table 4.1.

TABLE 4.1

Summary of Different Fields of Applications of Protein Engineering

Application Field	Engineered Product	References
Food industry	Enzymes such as proteases, amylase, cellulase, and xylanases	Kapoor et al. (2017)
Detergent industry	Proteases	
Environmental application	Peroxidase, laccases, and oxygenases	
Medical application	Insulin	
Biopolymer applications	Bacterial polymer such as polysaccharides, polyamides, and polyanhydrides	Chow et al. (2009)
Nanobiotechnology	Genetically engineered peptides	Howorka (2011)

4.3.1 Applications in Food

Protein engineering is a young discipline that has been stretched out from the field of hereditary building to the design of new enzymes. Enzyme biotechnology serves to improve the properties of enzymes such as thermal stability, specificity, and catalytic efficiency to improve their scope of application in food industries, that is, meat tenderization and cheese production. Different enzymes improved by DNA technology and protein engineering with practical application in food industry are amylases, proteases, lipases, phytases, and so on. Another important application of protein engineering in food industry is the engineering of gluten proteins.

In food industry, microbial proteases have been used in meat tenderization, milk clotting, flavor, and in low allergenic infant formulas. Different protein engineering techniques such as cold adaptation laboratory evolution of mesophilic subtilisin-like protease, DNA shuffling, site-directed mutagenesis, and/or random mutagenesis have been performed in protease-producing *Bacillus* species to improve activity at alkaline pH and extreme temperature conditions. Alpha amylases usually from *Bacillus* species have wide applications in food industry such as for starch hydrolysis to convert starch into glucose and fructose, for partial replacement of malt in brewery, and as flour improver in baking industry. Protein engineering techniques such as random mutagenesis, recombinant enzyme technology, and enzyme immobilization have been employed to improve enzyme activity, productivity, extreme temperature, and pH stability. Another important group of enzyme is lipases that serve the functions as dough conditioner and flavor enhancer in cheese industry. They have been engineered by lid swapping and DNA shuffling followed by computer modeling of lipase isoform to produce pure lipase from *Candida rugosa* (Turanli-yildiz et al. 2012).

One of the important applications of protein engineering in food industry is use of microorganism to produce gluten. Gluten protein expression requires expression system (*Escherichia coli*, *Saccharomyces cerevisiae*, *Pichia pastoris*, etc.), promoter and plasmid stability, and codon usage. Production of gluten through heterologous expression system has the advantages of high production, wide range of available vectors, use of fusion technology, cost efficient, and quicker production (Kapoor et al. 2017).

4.3.2 Medical Applications

Protein engineering is important for the development of the therapeutic protein with wide applications in different fields of medical science such as cancer treatment and antibody engineering. Protein engineering and recombinant DNA technology have increased the potential of radioimmunotherapy, the potential technique for cancer treatment, and the development of novel antibodies such as anticancer agents with high specificity to cancer cell. The progress made in protein engineering resulted in a second generation of therapeutic protein products with application-specific properties obtained by mutation and deletion of fusion. Protein engineering method such as DNA shuffling and genetic recombination helped in the development of biotherapeutic proteins, such as insulin, interferon, erythropoietin, and combinatorial protein with application in proteomics and therapeutics. Further, protein engineering has wide application in gene therapy, effective drug discovery, delivery, and tissue engineering (Turanli-yildiz et al. 2012). Development of insulin through mutagenesis to create monomeric forms or to precipitate upon injection is one of the examples of engineered therapeutic protein (Kapoor et al. 2017). Various new and innovative methodologies for repairing damaged myocardium are being studied. Protein engineering has growing scope in cardiovascular therapeutic development. Proteins engineered to induce cardiomyocyte proliferation and cardiac microvasculature formation have increased the potential of protein engineering in cardiac regeneration (Jay and Lee 2013).

Development of stimulus-responsive therapeutic proteins has been possible with recent advances in protein engineering. The certain advantages associated with engineered proteins include targeted drug neutralization, targeted drug activation, antibody–drug conjugates, and engineered zymogens that can act both as biosensor and effectors, directed enzyme prodrug therapy, and light responsive on/off control of drug activity (Tobin et al. 2014).

4.3.3 Environmental Applications

Microorganisms are being used for biotreatment and bioremediation of polluted environment, based on their expression of enzyme with specific catalytic activity toward waste materials. However, many synthetic compounds are resistant to biodegradation as microorganism lacks complete set of catabolic enzymes. With advances in protein engineering, catalytic activity of the enzymes can be modified by *in vitro* mutagenesis. Enzymes such as haloalkane dehalogenase, methane monooxygenase, cytochrome P450, ligninase, and aromatic deoxygenase, which have catalytic role for the degradation of pollutants such as halogens have been produced by utilizing engineered microbials (Janssen and Schanstra 1994). Pesticides, such as aldrin chlordane, industrial waste chemicals, such as dioxins and furans, and polychlorinated biphenyls are some of the examples of persistent organic pollutants that are resistant to environmental degradation leading to health-related problems in humans, aquatic, and animals. Some of the microorganisms are known to be capable of biodegrading these chemicals due to genetic adaptation to encode the biochemical mechanisms for dealing with potential substrates (Oyetibo et al. 2017).

4.3.4 Biopolymeric Nanomaterials and Their Functionalities

The ability of protein engineering to create and improve protein domains can be utilized for producing new biomaterials for medical and engineering applications. Protein engineering and macromolecular self-assembly together lead to the development of biopolymers. Engineered peptide-based biomaterials including poly-amino acids, elastin-like polypeptides, silk-like polymers, and other biopolymers are being focused for protein purification, controlled drug delivery, tissue engineering, and biosurface engineering applications (Chow et al. 2009).

Protein engineering broadens the scope of structural and functional repertoire of protein assemblies in nanobiotechnology. Repetitive nature and defined morphological shape of proteins can be exploited for different nano-biotechnology applications such as development of vaccine, biodelivery, biocatalysts, and material science. Engineered proteins can be utilized to enhance the effectiveness of vaccines by boosting the immune response and improving the enzyme catalytic activity. Engineered proteins act as nano-containers to concentrate nontethered enzymes, leading to improvement of biocatalytic activity. Nanoscale-engineered proteins can be used to create nanoparticles, such as silk threads with mechanical properties similar to natural template (Howorka 2011).

4.4 Proteomes and Proteomics

The term *proteome* is defined as the total population of proteins expressed in a cell at a particular time. Proteome expresses the status of cell, as proteome of diseased cell is different from that of the healthy cell. The proteome is principally dynamic and has an intrinsic convolution that exceeds genome or the mRNA complement found in the cell. The sequencing of several genomes has elevated the opportunities for the study of proteins. Marc Wilkins and his colleagues coined the term *proteome* and *proteomics* combining the terms *protein* and *genome* and therefore, proteomics refers to the study of proteome (Saraswathy and Ramalingam 2011). Proteomics is defined as the comprehensive identification and quantitative analysis of protein expression in cell line, tissue, organelle, or organisms at a specific time under certain conditions. It involves the study of functions, structure, composition, and interactions of the proteins that direct the activities of each living cell. Proteomics assists in understanding the modification in protein expression through different phases of the life cycle or under stress stipulations. It also helps in understanding the arrangement and function of different proteins as well as protein–protein interactions of an organism (Chandrasekhar et al. 2014).

The information regarding mechanism of disease, aging, and effect of the environment are lacking from the study of gene. Understanding of detailed proteomics may be helpful for identifying the causes and mechanisms. Proteomics helps to identify the total number of genes in particular genome, which is difficult to predict from the genomic data. To understand functional annotation of the genome, intron–exon structure of the gene, genomic, and proteomics need to be integrated. Proteomics helps to understand the correlation between mRNA and protein expression. During protein synthesis, mRNA is subjected to posttranscriptional control in the form of alternative splicing, polyadenylation, and mRNA editions, followed by translation of mRNA and posttranslational modification (PTM) of the proteins. PTM of protein in response to intracellular and extracellular stimuli can be inferred from proteomic analysis. Proteomics also helps to identify the protein function by examining their 3D structure. Subcellular location of protein can be identified by proteomics, which can be utilized for protein mapping, protein–protein interaction, and better understanding of protein regulation (Graves and Haystead 2002).

4.5 Genomics to Proteomics: Sequential Phenomena in Bionanotechnology

Human genome is estimated to have 30,000–40,000 genes that can encode around 40,000 different proteins. Due to alternate RNA splicing and PTMs, around 2,000,000 proteins can be encoded by human cells, which reflect that proteome is far more complex than the genome. Genomics provides information regarding gene activity related with disease but posttranslation information is lacking. On the other hand, the study of proteins introduces

PTMs and provides the information of the various biological functions. Gene sequence and gene activity do not provide complete and accurate information about protein final structure and activity because proteins get modified even after translation by PTM through the addition of carbohydrate and phosphate groups. These modifications are not directly coded by genes but are important in modulation protein function. Therefore DNA sequence does not predict the active form of protein, and RNA quantitation does not show the abundance of protein. Genome of particular cells remains relatively constant, whereas proteins may change as genes are turned on or off in response to the environment. As proteins are directly related to both normal and disease-associated biochemical processes, disease condition can be fully understood by directly analyzing the proteins (Cho 2007).

Figure 4.3 illustrates the process of formation of the functionally active structures of protein by undergoing cotranslational and PTM processes. When a gene is expressed, it results in transcription of coding DNA strand into mRNA, which is edited by intron elimination and joining of exons. A known mRNA sequence is capable of producing more than one protein. Although some properties of an organism can be associated with the activity of a single gene, mostly they are determined by the combined action of numerous gene products. The gene exerts its function only at the proteome level. The operation and the functions of living cells are usually an outcome of proteome dynamics. Protein–protein interactions are responsible for processing information (signal and regulatory proteins), maintenance of architectural characteristic (structural proteins), and regulation of cellular metabolism (enzymes). As proteome analysis provides a view of the biological processes at their level of occurrence, it offers an improved understanding of genomics of the cell cycle, cell death, development stage, cell function, and cellular responses to external stimuli and diseases. Proteomics has become a vital step in the development and validation of diagnostics and therapeutics (Cristea et al. 2012).

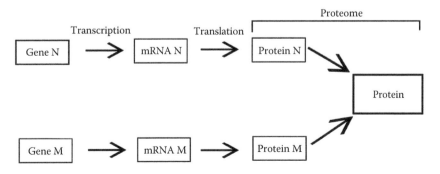

FIGURE 4.3
Succession from genome to proteome. (From Cristea, I.M. et al., *Blood*, 103, 3624–3634, 2012. With Permission.)

4.6 Classification of Proteomics

Broadly, proteomics is classified into seven branches based on the functions: structural proteomics, functional proteomics, quantitative proteomics, glycoproteomics, phosphoproteomics, expression proteomics, and clinical proteomics, dealing with high-throughput method for prediction of protein structure, protein function, protein quantification, glycoprotein analysis, phosphoprotein analysis, protein expression, and analysis of different proteins in biological samples, respectively (Saraswathy and Ramalingam 2011).

4.6.1 Structural Proteomics

Structural proteomics is related with the study of 3D structure and function of protein complex in specific cellular organelle with the help of techniques such as nuclear magnetic resonance (NMR) spectroscopy and X-ray crystallography. Structural proteomics therefore assists in the identification of proteins within a protein complex of organelles, determination of its location, and characterization of protein–protein interactions. The information obtained from this approach also helps to understand the mechanism how protein expression influences its structure and characteristics (Graves and Haystead 2002).

Structural proteomics helps to study the different levels of complexity in protein system, that is, single, binary, multisubunit protein complexes, proteome–protein interaction networks, organelles, cells, and tissues from which information about protein structure can be obtained. Information about amino acid residue level, degree of exposure of specific amino acid residue to solvent, and distance between functional groups can be determined from structural proteomics. This information can be used to identify the dynamic and conformational changes of the proteins or can be used as constraints for unknown protein molecular modeling (Petrotchenko and Borchers 2014).

4.6.2 Functional Proteomics

Functional proteomics is used to study and characterize the use of protein ligands and isolate specific types of proteins by affinity chromatography to gain information about the protein functions, protein–drug interactions, protein signaling, and disease mechanisms (Graves and Haystead 2002). *Functional proteomics* is a broad term from many approaches such as specific proteomics and directed proteomics. In specific proteomics, the target protein or protein ligand is isolated by affinity chromatography for analysis purpose. Isolation of specific protein and its characterization can provide

TABLE 4.2

Commonly Used Affinity Tags and Ligands in the
Isolation of Multiprotein Complexes

Tag	Ligand
Poly-His tail	Nickel ion
Biotin	Streptavidin
Calmodulin-binding peptide	Calmodulin (Ca++)
Glutathione S-transferase	Glutathione
Specific peptide epitope	Monoclonal antibody

Source: Adapted from Monti, M. et al., *Clin. Chimi. Acta.,* 357, 140–150, 2005.

information about protein signaling, disease mechanisms, or protein–drug interactions (Chandrasekhar et al. 2014). Some of the commonly utilized affinity tags and ligands in the isolation of multiprotein complexes are summarized in Table 4.2.

4.6.3 Quantitative Proteomics

Quantitative proteomics deal with relative quantitative measurement of protein expression between healthy versus diseased and treated versus control samples. Two-dimensional polyacrylamide gel electrophoresis (2D-PAGE) followed by mass spectrometry (MS) analysis, protein array-based, and MS signal intensity-based quantification methods are used for quantitative proteomics. Labeling of proteins by incorporating stable isotopes such as ^{15}N or ^{14}N nitrogen or stable isotope labeling with amino acids in culture (SILAC) such as ^{2}H-Leucine, ^{13}C-Lysine can be done to improve sensitivity of the protein detection. Interactome is the branch of proteomics dealing with the study of protein–protein interaction using methods such as protein affinity chromatography, immunoprecipitation, and phage display (Saraswathy and Ramalingam 2011).

4.6.4 Expression Proteomics

The expression proteomics evaluates the quantitative and qualitative expressions of proteins between samples that are different in some variables. In this approach, comparison can be performed between samples for protein expression of the entire proteome or subproteomes. For example, tumor tissue and normal tissues' protein expression can be analyzed by expression proteomics. Thus, novel proteins or disease-specific proteins can be identified from the information obtained from expression proteomics. MS and 2D gel electrophoresis are used to observe the difference in protein expression (Graves and Haystead 2002).

4.7 Technology of Proteomics

The strategic design for the isolation of protein targets is the most critical stage of proteomics. Due to advancement in mass spectroscopy, there has been an emphasis on the *front-end* of proteomic experiments compared to the data analysis, which can lead to the isolation of hundreds of irrelevant proteins for identification. Different techniques are employed for this and they involve three basic steps: (1) separation and isolation of protein, (2) acquisition of protein structural information for protein characterization and identification, and (3) database utilization (Graves and Haystead 2002).

4.7.1 Protein Separation and Isolation

For proteomics analysis, the protein of interest needs to be separated from other unwanted protein. The desired protein may be intracellular, membrane-bound, or extracellular. The extraction process may vary depending on location of protein, for example, membrane-bound or intracellular protein extraction requires disintegration of the cell, whereas extracellular protein extraction requires filtration or centrifugation steps to remove the unwanted biomass. For effective visualization, identification, and characterization of the protein, the desired protein needs to be separated from the protein mixtures (Saraswathy and Ramalingam 2011).

4.7.2 One-Dimensional Gel Electrophoresis

The predominant technology for protein separation and isolation is PAGE. One-dimensional gel electrophoresis or sodium dodecyl sulfate polyacrylamide gel electrophoresis (SDS-PAGE) involves protein separation based on molecular mass. In this method, amphoteric proteins are made to run in single direction with uniform negative charge created by SDS that even denatures the native protein structure by influencing the noncovalent forces such as hydrogen bonding, hydrophobic, and ionic interaction. The denatured linear proteins are loaded onto the polyacrylamide gel (PAGE), which has two phases: (1) a stacking gel and (2) separating gel. Under electric field, the stacking gel concentrates the SDS-loaded linear protein molecules, whereas the separating gel separates the denatured protein based on molecular weight. At the end, gel is stained by coomassie brilliant blue R250 dye and separated protein molecules are visualized under ultraviolet (UV) light. This is a simple reproducible method used to resolve proteins with molecular masses of 10–300 kDa (Saraswathy and Ramalingam 2011).

4.7.3 Two-Dimensional Gel Electrophoresis

In case of complex proteins, two-dimensional gel electrophoresis (2D-PAGE) is used where proteins are separated based on two properties, that is, net charge and molecular mass. With the progress achieved in this method, it is suitable for separating up to 10,000 proteins in a single gel analysis. One of the major advantages of 2D-PAGE is the ability to resolve protein that has undergone PTMs. This method serves in protein expression profiling where expression of two samples can be compared both qualitatively and quantitatively. It is also used in cell map proteomics for microorganisms, organelles, and protein complex (Graves and Haystead 2002).

4.7.4 Acquisition of Protein Structural Information

4.7.4.1 Edman Sequencing

The classical method for protein identification involves microsequencing by Edman chemistry. Edman sequencing technique was introduced in 1949 by Edman that identifies *N*-terminal sequence of amino acid. This method is utilized for the identification of protein separated by SDS-PAGE. The general procedure for mixed-peptide sequencing involves PAGE for the separation of protein mixture followed by transfer of the target protein to an inert membrane by electroblotting. During electroblotting, proteins are visualized on the membrane surface, and excised and fragmented into peptides. The membrane is placed into an automated Edman sequencer where 6–12 automated Edman cycles are carried out. The resulting sequence data are fed into the FASTF (protein database) or TFASTF (DNA database) algorithms, which sort and match the data against stored database and identify the target protein (Graves and Haystead 2002).

4.7.4.2 Mass Spectrometry

MS is one of the powerful techniques for analysis of biomolecules. This technique is useful in accurate protein identification and mass determination through peptide mass fingerprinting and *de novo* protein sequencing using tandem analysis. This method is also useful to identify posttranslation site modification such as phosphorylation and glycosylation. MS separates proteins and peptides based on their mass-to-charge ratio (m/z), and tandem mass spectrometry (MS/MS) fragment the parent protein molecule into daughter molecules (ions) and allow sequencing of the peptide constituents of protein. First, the molecules are ionized utilizing different ionization techniques such as electrospray ionization (ESI), matrix-assisted laser desorption/ionization (MALDI), and surface-enhanced laser desorption/ionization (SELDI). The ionized molecules are passed into a mass analyzer by an electric field. Under the electric current, ions are separated according to mass-to-charge ratio from where the information is passed by detector to the computer for analysis (Cho 2007). Two major tools used in quantitative proteomics are affinity chromatography such as enzyme-linked immunosorbent assay (ELISA) and MS. These tools when combined together as affinity-MS can help to measure low-abundance proteins with high specificity for quantitative proteomics (Li et al. 2017).

4.7.4.3 Bioinformatics and Proteomes

Data resulting from genomic and proteomic studies are analyzed by bio-informatics tool, which is an integration of mathematical, statistical, and computational methods to analyze biological, biochemical, and biophysical data. Proteomic database collection with biological information of proteins is utilized for annotation and interpretation of structural information obtained from Edman sequencing or MS. Thus, bioinformatics helps to convert raw proteomics data into information, for example, prediction of protein–protein interaction based on sequence information (Cho 2007).

Database searching aims to fasten the identification process and assist in accurate identification for large numbers of proteins. Different database processes such as peptide mass fingerprinting, amino acid sequence, and *de novo* peptide sequence are being utilized for identification of proteins. In peptide mass fingerprinting, masses of peptides from the proteolysis of target proteins are compared to the predicted masses of peptides from the theoretical digestion of proteins in a database. Amino acid sequence database searching employs identification of amino acid sequence by interpretation of the MS spectrum, which is further employed to identify unknown peptide sequence based on the molecular masses. *De novo* peptide sequence approach is based on identification of *de novo* sequence data from peptides by MS/MS and utilization of the peptide sequence to search appropriate database. This method is particularly suitable in the case that lacks well-annotated databases such as in *Xenopus laevis* or human. Different uninterpreted MS/MS data searching program such as Mascot, SONAR, and SEQUEST are also available for the identification of proteins (Graves and Haystead 2002).

4.8 Proteomics to Nanoproteomics

The progress made in nanotechnology over the past few decades has influenced the growth of different biological fields. Nanotechnology has widened the spectrum of different innovative techniques leading to an increase in the high throughput of existing proteomic approaches. Nanoproteomics is one of the innovative disciplines of nanobiotechnology resulting from the integration of nanotechnology and proteomics (biotechnology). Nanoproteomics was developed with the aim to overcome the limitation of existing proteomic techniques, for instance, immunoassays, protein microarrays, 2D-PAGE, and MS approaches, which have drawbacks such as limited dynamic range, poor sensitivity, complex separation and data analysis, and chances of false results. Nanoproteomics, therefore, helps in the development of robust analytical techniques that are fast, selective, and sensitive for low-abundance proteins and high-dynamic range proteins by enabling multiplexing and high-throughput analysis of low-volume proteins. It further helps in better understanding of proteomes and discovery of new biomarkers.

Initially, nanoproteomic techniques relied on label-based techniques that utilized nanobioreceptors such as enzymes, nucleic acids, antibodies, aptamer (synthetic functional oligonucleotide receptors), biomimetic receptors, microbial cells, or bacteriophages, whereas now emphasis is being given for the development of label-free nanoproteomics for biomarker discovery. Broadly, nanoproteomics is divided into label-free and label-based methods, which are based on optical, electrochemical, and mass-based methods. Miniaturized nanoproteomic platforms such as nano HPLC (high-performance liquid chromatography)-chip coupled with ion trap MS for the differential proteomics of lupin proteins, nanocapillary electrophoresis for analysis of different bioanalytes, capillary electromigration, microchip, and nano-LC (liquid chromatography)/capillary LC have been developed through integration of nanotechnology and biotechnology (Agrawal et al. 2013).

4.9 Applications of Nanoproteomics

The most important applications of proteomics are protein identification, molecular weight determination, isoelectric point determination, amino acid sequence, PTM, peptide mapping, protein structure determination, biomarker identification, and drug–target identification (Saraswathy and Ramalingam 2011). Proteomics has several practical applications in life sciences such as drug development against diseases, prediction of a gene from protein, development of personalized drugs, and analysis of the difference in protein expression profile between the diseased and normal person for a target protein. Once the protein is identified, its function can be predicted. Proteomics is also widely used in biological fields, being mainly applied in food microbiology, agriculture, oncology (tumor biology), and biomedicine. Studies are conducted on the application of proteomics on biofuel crops to manufacture proteomics-based fungicides and improve tolerance to abiotic and biotic stresses.

The information from proteomics can be used to identify proteins associated with a particular disease; this helps in the development of potential new drugs for their treatment. The pattern of change in protein expression induced by drugs helps to understand the mechanism of action of drugs. For introducing effective drugs, various drugs are grouped and compared based on their signaling cascades or metabolic pathways. The correlation of the dynamic expression of a proteome and the physiological changes related to diseased or healthy state helps to understand disease mechanism, validate disease models, discover biomarker, therapeutic targets, characterize drug effect, and study protein toxicology.

The process of spreading of cancer from one organ to another nonadjacent organ is referred as tumor metastasis, which can also be fatal for the patients. Understanding the protein expression of the metastatic process can help to recognize the molecular and cellular mechanisms underlying tumor metastasis. This would help in developing an approach for the clinical management of cancer and its therapeutic interference. Proteomics assist in identifying the functions of tumor cells and characterization of protein expressions, which are widely used in biomarker discovery (Chandrasekhar et al. 2014). Protein expression profiles are being used in cancer diagnostic and treatment. 2D analysis and protein characterization of bladder tumors showed correlation between protein expression patterns and squamous cell carcinomas (SCC), and psoriasin, a protein externalized to the urine, was identified as diagnostic biomarker. Similarly, 2D analysis of human heart proteins showed correlation between changes in myocardial protein expression of biopsies with heart disease. Protein profiling by 2DE showed calbindin as marker for cyclosporine-mediated nephrotoxicity in dogs, monkeys, and humans kidney tissues (Kellner 2000).

The application of proteomics in food technology helps in process development, quality control, detection of batch-to-batch variations, and standardization of raw materials. It also facilitates the research for microbial and biological safety, and use of genetically modified foods (Chandrasekhar et al. 2014). Analysis of proteome of food crops have variety of applications including determination of the nutritional values, improvement of the product quality by identifying protein markers and detection of the presence of food allergens. The authenticity of fruit and plant extracts used in the production of fruit juices can be done by using protein biomarkers (Fasoli et al. 2011).

Proteomics have been successfully applied over last decade for processing and safety of certain food such as wine and beer, meat, milk, and transgenic plant by exploratory analysis of food in parallel to the genomic and transcriptomic approaches. Health benefits associated with phytochemicals in food have been discovered in food proteomics, for example, soy-based diets rich in peptides, isoflavone, and genistein are found to stop cancer cell division in skin cancer, promote cardioprotective effect and antioxidative capacity, protect against atherosclerosis, and increase anti-inflammatory activity. Further, proteomic analysis of structural isomer-conjugated linoleic acid (CLA): trans-10-cis-12 CLA and cis-9-trans-11 CLA showed proatherogenic effect and antiatherogenic effect, respectively. Proteomic profiles of genetically modified organisms helped to detect unintended effects in transgenic crops. In addition, application of proteomics has led to the discovery and quantification of biomarkers that help to solve problems related to food quality, nutrition, and safety by assessing predisposition, efficacy, and characterization and quantification of bioactive food (Agrawal et al. 2013).

TABLE 4.3

Application of Proteomics for Food Authentication in Different Types of Foods

Food	Proteomic Technique	Purpose of Analysis
Milk	2-DE + PMF and MALDI-TOF MS	Adulteration of fresh milk with powder milk
Meat	2-DE	Differentiation of meat species
Shellfish	MALDI-TOF MS protein profiling library	Differentiation of shrimp species, origin, and fresh/frozen state
Fish	MS/MS spectral library	Differentiation of fish species
Wine	MALDI-TOF MS fingerprinting	Differentiation of wine varieties and addition of fining agents
Honey	MALDI-TOF protein profiling	Geographical origin
GMOs	2DE	Comparison of genetically modified and nonmodified maize, soybean, common bean, and potato

Source: Adapted from Ortea, I. et al., *J. Proteom.*, 147, 212–225, 2016.

One of the major applications of proteomic in food technology is assessment of food authentication regarding identification of breed, origin of food, adulteration of food stuff, types of adulterant, production and processing condition, ingredients added, and detection of genetically modified organisms (GMOs). The approaches to assess food authentication in different foods are summarized in Table 4.3.

The information of key proteins plays an important role in the best possible development and advancement of plant biotechnology. Research studies provide informational collections for product proteomics that reinforces the opportunities to have the capacity in which such proteomic-based learning is utilized specifically for the change of the anxiety resilience of a harvest plant (Agrawal et al. 2013).

Genomics and proteomics are two major disciplines for discovery of novel genes leading to improvement of crops. Two-dimensional electrophoresis (2-DE) and mass spectroscopy (MS) are two broadly utilized proteomics techniques that are utilized to recognize proteins in various proteome states or conditions. Proteomics offer novel genes (DNA) to induce improvements in crops, whereas marker-assisted selection (MAS) helps to choose gene of interest or quantitative characteristic loci (QTLs). Proteomics and MAS together are useful in breeding to produce offspring with different phenotypes. In plants, proteomics have wider application to improve their abiotic and biotic stress-tolerance mechanism. Proteomic approaches are useful to improve crop productivity via identification of disease-resistant gene, stress-tolerant gene, gene responsible for reduction in preharvest and postharvest loss and biomarkes responsible for optimal harvest maturity. (Eldakak et al. 2013).

References

Agrawal, G. K., A. M. Timperio, L. Zolla, V. Bansal, R. Shukla, and R. Rakwal. 2013. Biomarker discovery and applications for foods and beverages: Proteomics to nanoproteomics. *Journal of Proteomics* 20:74–92.

Antikainen, N. M. and S. F. Martin. 2005. Altering protein specificity: Techniques and applications. *Bioorganic and Medicinal Chemistry* 13:2701–2716.

Chandrasekhar, K., A. Dileep, D. E. Lebonah, and J. P. Kumari. 2014. A short review on proteomics and its applications. *International Letters of Natural Sciences* 17:77–84.

Chen, Z. and A. P. Zeng. 2016. Protein engineering approaches to chemical biotechnology. *Current Opinion in Biotechnology* 42:198–205.

Cho, W. C. S. 2007. Proteomics technologies and challenges. *Genomics, Proteomics and Bioinformatics* 5:77–85.

Chow, D., M. L. Nunalee, D. W. Lim, A. J. Simnick, and A. Chilkoti. 2009. Peptide-based biopolymers in biomedicine and biotechnology. *Materials Science and Engineering* 62:125–155.

Cristea, I. M., S. J. Gaskell, A. D. Whetton, and W. Dc. 2012. Proteomics techniques and their application to hematology. *Blood* 103:3624–3634.

Eldakak, M., S. I. M. Milad, A. I. Nawar, and J. S. Rohila. 2013. Proteomics: A biotechnology tool for crop improvement. *Frontiers in Plant Science* 4:1–12.

Fasoli, E., A. D'Amato, A. V. Kravchuk, A. Citterio, and P. G. Righetti. 2011. In-depth proteomic analysis of non-alcoholic beverages with peptide ligand libraries. I: Almond milk and orgeat syrup. *Journal of Proteomics* 74:1080–1090.

Galzie, Z. 1991. What is protein engineering? *Biochemistry and Molecular Biology Education*. doi:10.1016/0307-4412(91)90007-U.

Graves, P. R. and T. A. J. Haystead. 2002. Molecular biologist's guide to proteomics. *Microbiology and Molecular Biology Reviews* 66:39–63.

Howorka, S. 2011. Rationally engineering natural protein assemblies in nanobiotechnology. *Current Opinion in Biotechnology* 22:485–491.

Janssen, D. B. and J. P. Schanstra. 1994. Engineering proteins for environmental applications. *Current Opinion in Biotechnology* 5:253–259.

Jay, S. M. and R. T. Lee. 2013. Protein engineering for cardiovascular therapeutics: Untapped potential for cardia repair. *National Institute of Health* 113:933–944.

Kapoor, S., A. Rafiq, and S. Sharma. 2017. Protein engineering and its applications in food industry. *Critical Reviews in Food Science and Nutrition* 57:2321–2329.

Kellner, R. 2000. Proteomics concepts and perspectives. *Fresenius' Journal of Analytical Chemistry* 366:517–524.

Li, H., R. Popp, and C. H. Borchers. 2017. Affinity-mass spectrometric technologies for quantitative proteomics in biological fluids. *TrAC Trends in Analytical Chemistry* 90:80–88.

Lutz, S. 2010. Beyond directed evolution-semi-rational protein engineering and desig. *Current Opinion in Biotechnology* 21:734–743.

Mandell, D. J. and T. Kortemme. 2009. Backbone flexibility in computational protein design. *Current Opinion in Biotechnology* 20:420–428.

Monti, M., S. Orrù, D. Pagnozzi, and P. Pucci. 2005. Functional proteomics. *Clinica Chimica Acta* 357:140–150.

Ortea, I., G. O'Connor, and A. Maquet. 2016. Review on proteomics for food authentication. *Journal of Proteomics* 147:212–225.

Oyetibo, G. O., K. Miyauchi, Y. Huang et al. 2017. Biotechnological remedies for the estuarine environmet polluted with heavy metals and persistent organic pollutants. *International Biodeterioration and Biodegradation* 119:614–625.

Petrotchenko, E. V. and C. H. Borchers. 2014. Modern mass spectrometry-based structural proteomics. *Advances in Protein Chemistry and Structural Biology* 95:193–213.

Saraswathy, N. and P. Ramalingam. 2011. *Concepts and Techniques in Genomics and Proteomics*. Oxford, UK: Woodhead Publishing.

Tobin, P. H., D. H. Richards, R. A. Callender, and C. J. Wilson. 2014. Protein engineering: A new frontier for biological therapeutics. *Current Drug Metabolisum* 15:743–756.

Turanli-Yildiz, B., C. Alkim, and Z. P. Cakar. 2012. Protein engineering methods and applications. *In Tech*, 978:953–51

Verma, R., U. Schwaneberg, and D. Roccatano. 2012. Computer-aided protein directed evolution: A review of web servers, databases and other computational tools for protein engineering. *Computational and Structural Biotechnology Journal* 2(3):e201209008.

5

Immune Systems, Molecular Diagnostics, and Bionanotechnology

5.1 Introduction

Emil von Behring, the father of immunology, and Shibasaburo Kitasato discovered a compound in blood that neutralized diphtheria toxin, which was later termed as *Antikörper* or *antibodies*. Antibodies were observed to show toxin specificity, that is, the ability to distinguish between two immune components. The terms *Antisomatogen* and *Immunkörperbildner* were coined referring to the substance that caused the formation of antibody, which gave rise to the term *antigen*. Hence, antibody and its cognate antigen led to the foundation of immunology (Schroeder and Cavacini 2010).

Innate immunity and adaptive immunity are the two types of human immune responses. Innate immunity system, including components such as epithelial cells, phagocytic cells, cytokines, natural killer cells, and complement system, is the early defense system for the development of adaptive immunity (Moulds 2009). Adaptive or humoral immunity shows immune response against foreign materials and pathogens via cell-mediated and antibody-mediated response mechanisms. In cell-mediated response, T-lymphocytes cells are activated to eliminate pathogens that exist within host cell and in antibody-mediated response and B-lymphocyte cells are activated to eliminate pathogens surviving outside the host cells (Moticka 2016).

5.2 Antibodies

Antibodies, also known as immunoglobulins (Ig), are glycoproteins produced by B-lymphocytes in response to antigens and foreign body particles (e.g., virus, bacteria, endoparasite, drugs, dust, and pollen grains). During adaptive immunity response, antigen-specific B-lymphocyte recognizes antigens with the help of antibody (B-cell receptor), which are activated with the assistance of helper T-lymphocytes (Th cell). The Th cells, in turn, are

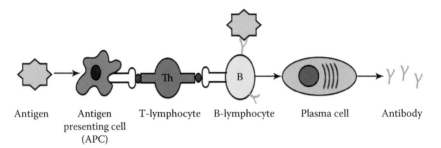

Antigen   Antigen   T-lymphocyte   B-lymphocyte   Plasma cell   Antibody
presenting cell
(APC)

FIGURE 5.1
Production of antibody. (From Day, M.J., *Top. Companion Anim. Med.*, 30, 128–131, 2015.)

activated by antigen-presenting cells (APC) displaying a processed form of antigens. With molecular interaction between B-cells and Th cells, Th cells produce cytokines that bind specific cytokine receptor expressed by the B-cells. Henceforth, B-cells activated by antigens divide and transform into plasma cells that produce antigen-specific Ig. Ig is divided into five classes (IgA, IgD, IgE, IgG, and IgM) and the type of Ig secreted depends on the response, for instance, IgG, IgA, and IgE are produced on response to systemic infections, mucosal pathogens, allergens, or parasites, respectively (Day 2015). The general scheme of the production of the antibody in response to the antigen is illustrated in Figure 5.1.

5.3 Antibody Structure

Antibodies are composed of Y-shaped monomeric subunits composed of pair of heavy (H) chains of size around 50–75 kDa and each heavy chain paired with one of a pair of identical light (L) chains of size around 25 kDa. The light chains exist either in κ or λ isoform, whereas the heavy chains exist in different isoform, which in turn regulates the class of Ig, such that α, δ, ε, γ, and μ sequence define the class IgA, IgD, IgE, IgG, and IgM, respectively (Strohl and Strohl 2012).

In IgG, both H-chain and L-chain are organized into domains of around 110–130 amino acids as follows:

    a. One NH_2-terminal variable (V) domain, V_L of L-chain, and V_H of H-chain

    b. COOH-terminal constant (C) domain, such that L-chain has one (C_L) H-chain has 3°C domain (C_H1, C_H2, C_H3).

Each domain consists of two sandwiched β-pleated sheets joined by disulfide bond between two cysteine residues. H chain with 3C domain has a

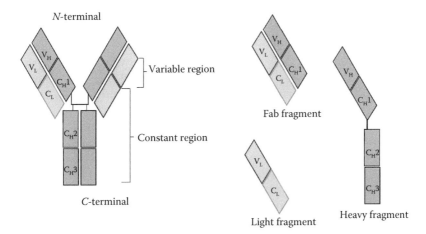

FIGURE 5.2
Structure of immunoglobulin. (From Ndoja, S. and Lima, H., Monoclonal antibodies. In *Current Developments in Biotechnology and Bioengineering* (Eds.) V. Thomaz-Soccol, A. Pandey, and R.R. Resende, pp. 71–95. Tokyo, Japan: Elsevier, 2017. With Permission.)

spacer hinge between the first and second domains, that is, between C_H1 and C_H2 (Schroeder and Cavacini 2010). The schematic representation of the antibody structure is shown in Figure 5.2.

On controlled digestion of chains with proteolytic enzymes, two fragments can be produced, that is, fragment for antigen binding (Fab) and fragment crystallizable (Fc). The Fc domains, which regulate cell killing and half-life of plasma mechanisms, consist of a pair of hinge: C_H2 and C_H3 (Strohl and Strohl 2012). Fab domain, which shows affinity and specificity to antigens, comprises both heavy and light chain components. It consists of entire L-chain (including both V_L and C_L portion) and V_H and C_H1 portion of H-chain. Fab domain is further divided into variable fragment (Fv) that comprises V_H and V_L domains and constant fragment that includes C_L and C_H1 domains (Schroeder and Cavacini 2010). Fab fragment can be obtained either by recombinant synthesis or by proteolytic breakdown of native antibody and the fragments with disulfide bridge thiols are known as Fab' fragments, whereas those lacking a thiol group are known as Fab fragments. Fab' fragments with thiol groups assist in their application in biosensor (Crivianu-Gaita and Thompson 2016).

5.4 Monoclonal Antibodies

Most of the antibodies exhibit polyclonal immune response, which make them heterogeneous in nature and difficult to characterize. Monoclonal antibodies (mAbs), on the other hand, are highly specific and sensitive. Monoclonal and polyclonal antibodies are similar in structure but vary in clone of the cells that produce them. mAbs are produced by a single clone, whereas polyclonal antibodies are produced by numerous clones (Singh et al. 2013). In 1975, Kohler and Milstein discovered the cell-hybridization technology, which finally led to the production of mAbs and the possibility to distinguish between healthy and unhealthy cells. The first mAbs were produced against sheep red blood cells (RBC) and this development revolutionized the concept of cell-surface components and led to the identification of tumor-associated antigens (Bernareggi et al. 2014).

The production involves three types of cells: (1) myeloma, (2) hybridomas, and (3) splenocytes as shown in Figure 5.3. Splenocytes are extracted and fused with myeloma cells using polyethylene glycol (PEG) as an agent to fuse the adjacent plasma membrane. The cells are cultured in hypoxanthine, aminopterin, and thymidine (HAT) media. In the HAT culture medium, myeloma cells die because they lack the hypoxanthine–guanine phosphoribosyl transferase (HGPRT) gene, which is required for the nucleotide synthesis and cell division. Splenocytes cells are also unable to grow and they eventually die. On account of this, only hybridomas are left in the media that can multiply indefinitely leading to rapid formation of colonies. The hybridoma cells are tested to ensure specificity and they are cultured in one medium from a single-cell colony to uniformity of antibodies. Thus, hybridoma technology leads to the mass production of mAbs

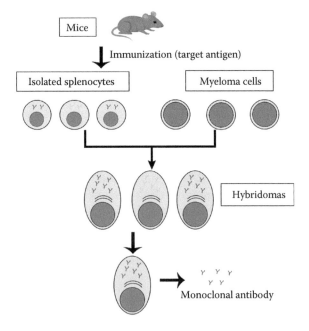

FIGURE 5.3
Steps in production of monoclonal antibodies. (From Ndoja, S. and Lima, H., Monoclonal antibodies. In *Current Developments in Biotechnology and Bioengineering* (Eds.) V. Thomaz-Soccol, A. Pandey, and R.R. Resende, pp. 71–95. Tokyo, Japan: Elsevier, 2017. With Permission.)

specific to pathogens or tumor cells. However, as these mAbs originated from mouse, human immune system recognizes them as antigens and produces human anti-mouse antibodies. With the advancement achieved in genetic engineering, it is now possible to produce mAbs using human genomic sequence (Ndoja and Lima 2017).

5.5 Recombinant Antibodies

With the advent of complementary DNA cloning, polymerase chain reactions (PCRs) have facilitated identification of the antibodies' genes and their expression in the recombinant host have led to the development of recombinant antibodies, single-chain antibody fragment (scFv). scFv is composed of L-chain variable domain (L_H) of antibody fused to H-chain variable domain (H_V) via a short peptide linker (Alvarenga et al. 2017).

The popularity of a recombinant antibody in medical and biomedical discipline is increasing because of the associated specificity, sensitivity, and stability. For the production of recombinant antibodies with greater human immune response, chickens are considered to be better host than other rodents and rabbits because of large phylogenetic distance between human and chicken. Further, chicken antibodies have only functional copy of variable segment of H-chain (V_H) and L-chain (V_L), which makes recombinant antibody generation easier. During recombinant antibodies production, initially host is immunized with the target antigen and boosted (3–4 times) followed by checking of serum antibody titer after each boost. When the desired serum antibody titer is reached, RNA is extracted from the bone marrow and spleen of the host, and finally cDNA is synthesized. V_H and V_L domains are amplified utilizing cDNA as template followed by joining of these segments by splicing by overlapping extension-polymerase chain reaction (SOE-PCR) to produce scFv. Thus, the produced scFv is cloned with the phage display vector and then transformed into *Escherichia coli* for antibody expression (Ma and O'Kennedy 2017).

5.6 Antigen–Antibody Interaction

Antibodies serve the immune response against antigens by different mechanisms, which include the following: (a) neutralizing biological activity or toxic metabolites of microorganisms; (b) activating pathways that cause bacterial cell lysis and inflammation; (c) enhancing uptake of antigens by phagocytic cells (opsonization); (d) involving in antibody-dependent cell-mediated cytotoxicity; and (e) releasing vasoactive mediators, histamine by degranulation of mast cells, and eosinophils (Moticka 2016).

Antigen–antibody reaction has been classified as follows:

1. Primary reaction, which involves formation of antigen–antibody complex formation by binding of an antigen to an antibody via an antigen-binding fragment.
2. Secondary reaction involves visible results such as agglutination, precipitation, flocculation, and so on, resulting from antigen–antibody complex.
3. Tertiary reaction involves *in vivo* biological manifestations of the reactions (Cruse and Lewis 1999).

Antigen–antibody interaction occurs between the epitope, small restrict particles on the surface of antigens and paratope, which is the site on the antibody where the antigen binds. Antibodies synthesized against antigens can identify surface epitopes, which represent conformational structures. On account of this ability, antibodies exhibit specificity, and therefore can differentiate between two closely related antigens or can bind to divergent antigens with similar epitopes (cross-reactivity) (Schroeder and Cavacini 2010).

Energy is required for the formation antigen–antibody complex, which also regulates the molecular specificity of antibodies. The binding strength of paratope of an antibody with the epitope of specific antigen is known as antibody affinity, which in turn is the sum total of attractive force and repulsive force. The binding strength required for binding between all sites of multivalent antigen and antibody is known as avidity, and hence is higher than affinity (Božič et al. 2014). In case of high affinity, epitope fits exactly in the paratope (lock and key interaction), whereas in case of low affinity, epitopes are loosely held in the paratope (Day 2015). Small portion of molecules involving few amino acids and small surface area between 0.4 and 8 nm^2 are involved in antigen–antibody interactions, which also have to overcome repulsive force. When antigens and antibodies are at several nanometer distances, they come together by attractive forces such as ionic bond, hydrophobic bond, and overcome the hydration energies by removing water molecules and come even more closer because of van der Waals force along with ionic strength. The binding strength of antigen–antibody complex at this condition depends on the contact area and goodness of fit between surfaces (Reverberi and Reverberi 2007).

5.7 Nanoparticles–Antibodies Bioconjugation

Variety of nanoparticles (quantum dots, nanotubes, silver, gold, silica, and magnetic nanoparticles) conjugated with biomolecules (antibodies, aptamer, peptide, and carbohydrate) have improved the scope of nanomaterials in the field of nanomedicine, biomedical nanotechnology, nanobiosensors, diagnostic, and drug delivery. Materials at nanoscale ranges are found to exhibit greater light-scattering properties and produce photobleaching-resistant optical signals compared to counterparts. Noble metallic nanoparticles (gold and silver) have the ability to scatter and absorb light strongly in the visible region because of the presence of surface plasmon resonances (SPR). As the nanoparticles exhibit better optical, electronic, and catalytic properties along with biocompatibility, they have been extensively used in an analytical biosensor (Shanker et al. 2014).

Bioconjugation is the process of bonding biologically active molecules to nanoparticles either by biological or chemical process for enhancing the stability, functionality, and biocompatibility. The physical means of conjugating gold nanoparticles (AuNPs) with biomolecules such as monoclonal antibody include three different mechanisms, that is, (1) ionic interaction, (2) hydrophobic interaction, and (3) dative binding. Chemical bioconjugation method involves the following: (1) chemisorption, (2) bifunctional linker, and (3) adapter. For bioconjugation, both covalent and noncovalent immobilization modes are being used. During electrostatic and hydrophobic interaction, antibodies are nonspecifically adsorbed onto gold (noncovalent mode), whereas during dative binding and chemical bioconjugation method, antibodies bind onto the surface of AuNPs by covalent mode. The different mechanism for bioconjugation of antibodies and nanoparticles is summarized in Table 5.1.

TABLE 5.1

Immobilization Mode and Method of Bioconjugation between Antibodies and Gold Nanoparticles

Immobilization Mode	Bioconjugation Method	Detail
Noncovalent	Ionic interaction	Negatively charged gold particles and positively charged group of antibodies interact for bioconjugation.
	Hydrophobic interaction	Interaction between hydrophobic groups of gold and antibodies.
Covalent bond	Dative binding	Interaction between the electron of gold nanoparticles (AuNPs) and sulfur atom of amino acid in antibody.
Covalent bond	Chemisorption	Interaction between gold and antibody through thiol group derivatives.
	Bifunctional linker	N-hydroxy-succinimide, (NHS), 1-Ethyl-3-[3-dimethylaminopropyl] carbodiimide (EDC) used to form amide bonds.
	Adapter molecules	Avidin and biotin are used to form the complex.

Source: Adapted from Jazayeri, M.H. et al., *Sen. Bio-Sen. Res.*, 9, 17–22, 2016.

5.8 Therapeutic Applications of Antibody-Based Bioconjugates

Target drug delivery and tissue engineering are the major therapeutic applications of antibody-based bioconjugates. mAbs and recombinant antibodies themselves possess pharmacological qualities. Three types of antibody-based bioconjugates: (1) antibody-drug conjugates (ADCs), (2) antibody-biologic conjugates (ABCs), and (3) antibody-nanoparticle conjugates (ANCs) with the size ranging from 10 to 100 nm, have widened the scope of therapeutic applications of bioconjugates. ADCs, composed of antibody bonded to drugs via linker, combine the cytotoxic property of drugs and targeting properties of antibody, resulting in an enhanced potency of antibody and reduced liver toxicity. Antibody of human origin used in ADCs helps to minimize human immunogenicity. ABCs composed of antibody with other biologics (protein, nucleic acids, and mAbs) exhibit functions similar to antibodies. ANCs combine the therapeutics properties of antibody and diverse properties of nanoparticles such as large payload, intracellular delivery, target drug delivery, and controlled release resulting in greater targeting properties and multifunctionality (Kennedy et al. 2017).

Two mAbs, one functioning to pass the conjugate across the blood-tumor barrier and other to target the tumor cell surface-bound nucleosomes, conjugated with the same polymaleic acid-based nanoparticle is one of the examples of target delivery of the bioconjugates. Antibodies or Fab fragments-functionalized nanoparticles have been utilized as a carrier of plasmid DNA and oligonucleotides to transfect cells and produce therapeutic proteins. Antibody–AuNP bioconjugates have been found to optically absorb nanoparticles for enhanced tissue repairing (Arruebo et al. 2009).

5.9 Diagnostic Applications of Antibody-Based Bioconjugates

Diagnostic field of applications of antibody-based bioconjugates can be either *in vivo* or *in vitro*, including cell sorting, bioseparation, imaging, enzyme immobilization, immunoassays, cell sorting, transfection, and biosensor. Cell sorting and bioseparation, utilizing antibodies attached to magnetic nanoparticles, are the basis of immunoseparation. Murine mAbs conjugated with magnetic beads were used for the selective selection of tumor cells. Utilization of antibody-conjugated nanoparticles in different imaging techniques such as magnetic resonance imaging (MRI), positron emission tomography (PET), computed tomography (CT), ultrasound, radiography, and so on, improved the selectivity and sensitivity of these technologies. Application of nanoparticles in different diagnostic approaches such as enzyme-linked immunosorbent assay (ELISA) offers the physical properties or can act as carrier of antibodies to recognize them by association in biosensor. Surface plasmon resonance (SPR) biosensor was used to detect the interaction between mAbs and nanoparticles (Arruebo et al. 2009).

5.10 Antibodies-Based Bioconjugates in Biosensor

Biosensors are the analytical machines comprising three utmost important parts: biorecognition system and transducer coupled with signal processor (Figure 5.4). One of the most important factors to be considered during biosensor fabrication is the selection of suitable biorecognition elements for rapid and reliable results (Mazur et al. 2017). The biological recognition element may be an enzyme, microorganism, tissue, or bioligand (antibodies and nucleic acids) and the transduction may be optical, electrochemical, thermometric, piezoelectric, magnetic, and micromechanical or combinations of one or more of the above-mentioned techniques (Velusamy et al. 2010).

The bioreceptor recognizes the target analyte and the corresponding biological responses are then converted into equivalent electrical signals by the transducer. The amplifier in the biosensor responds to the small input signal from the transducer and delivers a large output signal that contains the essential waveform features of an input signal. The amplified signal is

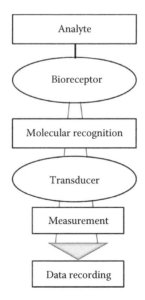

FIGURE 5.4
Schematic diagram of biosensor. (From Vo-Dinh, T. and Cullum, B., *Fresen. J. Anal. Chem.*, 366, 540–551, 2000. With Permission.)

then processed by the signal processor where it can later be stored, displayed, and analyzed. Biosensors have been widely applied to a variety of analytical problems in medicine, environment, food processing industries, and security and defense mechanism (Taylor et al. 2006; Velusamy et al. 2010).

In general, antibody/antigen, enzymes, nucleic acids, cells/cellular structures, biomimetic, and bacteriophage are the commonly utilized bioreceptors. The enzymes, antibodies, and nucleic acids are the main classes of bioreceptors, which are widely used in biosensor applications. Though enzymes are one of the biorecognition elements, they are mostly used to function as labels than the actual bioreceptor (Roda et al. 2012).

There are different types of biosensors utilizing antibodies as immobilized probe (ligand): mass-based biosensors, optical biosensors, and electrochemical biosensors, which are used to analyze drug components, biomarkers, pathogens, toxins, or pesticides. Biosensors are classified into two main categories: (1) biocatalytic biosensors, which involve immobilized cells, organelles, and enzymes such that interactions between ligand and analyte induce chemical changes and (2) affinity biosensors, which involve immobilized antibody, antigen, aptamer, DNA, and membranous receptor such that interactions between the ligand and analyte do not induce chemical changes (Patris et al. 2016).

5.10.1 Antibodies as Bioreceptors

Antibodies are common bioreceptors used in biosensors. Antibodies may be polyclonal, monoclonal, or recombinant, depending on their selective properties and the way they are synthesized. In any case, they are generally immobilized on a substrate, which can be the detector surface, its vicinity, or a carrier. The basic structure of an antibody and antigen–antibody lock is illustrated in Figure 5.5.

The way in which an antigen and an antigen-specific antibody interact is similar to a lock and key fit. An antigen-specific antibody fits its unique antigen in a highly specific manner, so that the three-dimensional (3D) structure of antigen and antibody molecules are matching. On account of this 3D shape fitting and the diversity inherent in individual antibody makeup, it is possible to find an antibody that can recognize and bind to any one of a large variety of molecular shapes (Meulenberg 2012). This unique property of antibodies is the key that makes the immunosensors a powerful analytical tool and their ability to recognize molecular structures that allow developing antibodies that bind specifically to chemicals, biomolecules, microorganisms, and so on. One can then use such antibodies as specific probes to recognize and bind to the analyte of interest that is present, even in extremely small amounts, within a large number of other chemical substances (Vo-Dinh et al. 2000).

Among the different classes of Ig, IgG, mAbs, scFv, Fab', and aptamers are the most dominant biorecognition elements used in biosensors. The size of scFv fragment and Fab' fragments is smaller than the whole antibodies (around 10–15 nm), which permits higher density immobilization of these

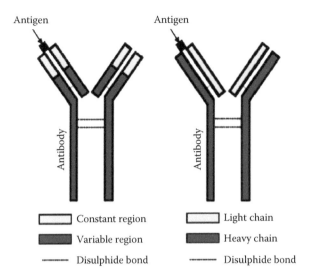

FIGURE 5.5
The basic structure of an antibody–antigen lock. (From Vo-Dinh, T. and Cullum, B., *Fresen. J. Anal. Chem.*, 366, 540–551, 2000. With Permission.)

fragments onto surfaces leading biosensors with higher sensitivity and lower limits of detection (LOD). Both scFv and Fab' fragments have been used as biorecognition elements in piezoelectric biosensors (quartz crystal microbalances [QCM]) and optical biosensors. scFv fragments have been prominently used in immunoassays, whereas Fab' fragments have wider applications in electrochemical biosensors (Crivianu-Gaita and Thompson 2016).

5.10.2 Immunoassay Measurement Formats

There are two fundamental immunoassay techniques: (1) heterogeneous and (2) homogenous. In heterogeneous immunoassays, an antibody (or antigen) is immobilized on a transducer support and the binding interaction with an analyte taking place at the interface is followed by monitoring the sensor response. In case of homogeneous assays, the biochemical reaction takes place entirely in the solution phase (Hodnik and Anderluh 2009).

Immunoassay mechanisms are further classified by the choice of the methodology of detection, which greatly depends on the nature of the target analyte, analytical sample, sensitivity of the analytical instrument, and application. The most frequently used measurement formats are as follows:

- Direct detection
- Sandwich assay
- Displacement assay
- Indirect competitive inhibition assay

5.10.2.1 Direct Detection Immunoassay Method

Direct immunoassay is done by immobilization of antibodies on the sensor surface and subjected to interact with the analyte of interest (Figure 5.6). The binding interaction of antibodies and the analyte causes the change in resonance angle and is directly proportional to the concentration of the analyte (Meulenberg 2012). It is useful for the detection of large molecules having molecular weight (MW) >10 kDa, because small molecules have insufficient mass to effect a measurable change in the refractive index. By the use of such simple direct immunoassay, Sonezaki et al. (2009) have demonstrated a linear dose–response curve for hemoglobin in the concentration range of 50 ppb–20 ppm with the covalently bound antibody on the carboxymethyl dextran surface.

5.10.2.2 Sandwich Assay

Sandwich assay is done in two recognition steps. In the first step, an antibody immobilized on a transducer surface is allowed for binding with an analyte of interest, whereas in the second step, a secondary antibody is left to flow over the sensor surface to bind with the previously captured analyte (Figure 5.7). This improves the sensitivity and specificity. This method is also suitable for only large molecules (Jang et al. 2009).

In antibody-based biosensors working on *sandwich* mechanism, antibody corresponding to a target antigen is immobilized onto the surface, followed by the binding of analyte to antibody. The immobilized analytes then bind to the second antibody forming a complex structure, which can detect the analytes by a marker attached to the second antibody. On the other hand, label-free biosensors are built with the antibody immobilized onto nanomaterials, such as piezoelectric biosensors, electrochemical biosensors, SPR biosensors, and so on, that can measure change in mass, electric current, and refractive index, respectively, resulting from the binding of the analyte to its specific antibodies (Burlage and Tillmann 2017).

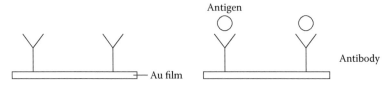

FIGURE 5.6
Schematic view of direct immunoassay. (From Li, Y. et al., *Food Chem.*, 132, 1549–1554, 2012. With Permission.)

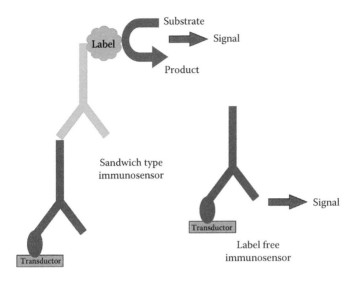

FIGURE 5.7
Schematic representation of sandwich type and label-free immunosensor. (From Patris, S. et al., *Trends Anal. Chem.*, 79, 239–246, 2016. With Permission.)

5.10.2.3 Indirect Competitive Inhibition Assay

Indirect competitive inhibition assay is used for the detection of low molecular weight molecules such as toxins, hormones, and chemicals. Mycotoxin detection is usually done with this assay in SPR biosensor in which antibodies act as a biological recognition element.

When mycotoxin conjugated with protein (e.g., bovine serum albumin [BSA] and ovalbumin), or mycotoxin derivatives were preimmobilized onto the sensor chip, there was competition between the preimmobilized mycotoxin and free mycotoxin in the sample toward the mycotoxin-specific antibody. If there is no mycotoxin in the sample, the antibodies will bind to the mycotoxin anchored on the sensor chip, leading to a larger value in response units (RU). On the other hand, the competitive reaction between mycotoxin in the sample and the anchored mycotoxin will lead to a lower RU value. With the increase in the concentration of mycotoxin in the sample, the extent of inhibition will be higher; thus, a much lower RU value will be obtained. Therefore, the amount of antibodies that bind to the chip surface is inversely proportional to the concentration of mycotoxin in the sample. Till now, the competitive inhibition assay has attracted much attention because of its high sensitivity on detection of small molecules (Li et al. 2012).

5.11 Types of Antibody-Based Biosensors

5.11.1 Optical Biosensors

Nanoparticles, such as gold that exhibits UV-visible absorption band, on SPR region show properties for better spectroscopy signals such as absorption selective to wavelength, large electromagnetics field, and molar extinction coefficient. AuNPs have been used in optical, multiphoton plasmon resonance, optical coherence, and third harmonic microscopy. Antibodies conjugated to AuNPs have been used for protein detection with great sensibility (sub-nanograms/mL) using infrared spectroscopy. Similarly, goat anti-mouse antibodies conjugated with gold-plated silver nanoparticles have been used in surface-enhanced Raman spectroscopy (SER) (Arruebo et al. 2009). Different research works have been conducted on optical biosensor utilizing antibody-based bioconjugates such as anti-insulin Fab' fragments immobilized onto AuNPs and Fab' fragments conjugated with 3,4-methylenedioxymethamphetamine (MDMA) on gold (Crivianu-Gaita and Thompson 2016).

SPR is a modern analysis technique based on the changes in the refractive index of material on the metal surface. In 1982, Liedberg first realized the biosensing potential of a prism SPR sensor with an IgG antibody adsorbed on the gold-sensing film, which allowed selective binding detection of IgG (Lyon et al. 1999). SPR was first demonstrated by Otto in 1968, but was not made commercially available for biomolecular interaction applications until the fall of 1990 by Biacore® (GE Healthcare). Originally, the SPR technique was applied in the analysis of gases, liquids, and solids (Hodnik and Anderluh 2009).

SPR is a physical optical phenomenon based on the change in the refractive index on the metal surface. A plane-polarized light beam entering the higher refractive index medium (glass prism) can undergo total internal refraction above a critical angle of incidence. Under these conditions, an electromagnetic field light component, that is, evanescent wave will penetrate into the gold film. At a specific angle of incidence, interaction of this wave with free oscillating electrons at the gold film surface will cause the excitation of surface plasmon's resulting subsequently in a decrease in the reflected light intensity. This phenomenon is called SPR and occurs only at a specific angle of incident light. SPR system thus detects changes in the refractive index of the surface layer of a solution in contact with the sensor chip (Hodnik and Anderluh 2009).

At a selected incident light wavelength or angle, the evanescent waves can resonate with surface plasmons (SP) produced by free electrons on the metal film of the sensor surface, and the energy of incident light will be absorbed by the SP as a sharp dip in the reflected intensity at a certain angle, which is dependent on the refractive index of medium on the nonilluminated side of the surface. The angle at which the minimum intensity of the reflected light is achieved is called SPR angle and this angel shifts when biomolecules bind to the surface and change the refractive index of the surface layer.

The resonance conditions are influenced by the biomolecules immobilized on the gold layer. Adsorption of biomolecules (antigen or antibody) on the metallic film, any followed conformational changes of the adsorbed biomolecules (or subsequent modification), and molecular interactions with relevant substances can be accurately detected. When a surface immobilized antibody binds with an analyte, the change in the interfacial refractive index can be detected as a shift in the resonance angle (Shankaran et al. 2007).

These changes are monitored over time and converted into a sensogram. The sensogram is a plot of SPR angle against time, and displays the real-time progress of the interaction at the sensor surface (Figure 5.6). Most of the SPR biosensors utilize RU. The signal is proportional to the amount of the bound molecule. For proteins, it was estimated that approximately 1000 RU corresponds to a surface coverage of 1 ng/mL unit, which indicates the change in reflection intensity with respect to incident angle before and after binding of the target molecule (Stenberg et al. 1991).

5.11.2 Mass Biosensor

QCMs and cantilevers are the two widely used mass biosensors. The shifting of piezoelectric crystal's frequency is directly proportional to the adsorbed molecular mass, which is the working principle of QCM. In one of the reverse QCM biosensor involving scFv fragments, parathion (analyte) was immobilized onto gold-coated quartz and anti-parathion scFv fragments were attached to parathion. However, the common QCM biosensor involves direct immobilization of antibodies followed by detection of analytes, for instance, anti-human IgG Fab' fragments conjugated onto AuNPs and analyte detected through QCM platform (Crivianu-Gaita and Thompson 2016).

5.11.3 Electrochemical Biosensors

On the basis of transduction process, potentiometric, amperometric, and conductometric are the common types of electrochemical biosensors, which detect the analyte on the basis of the electrochemical stripping of nanoparticles. Electrochemical biosensors utilize specific biorecognition elements such as oligonucleotides probes, immobilized antibodies, and so on. Analytes are detected in a sandwich format by the analysis of reaction of trapped analytes with the secondary antibody conjugated with electroactive nanoparticles (Arruebo et al. 2009).

5.12 Antibody-Based Biosensor for Detection of Pathogens

Pathogens (bacteria such as *Bacillus spp., Campylobacter spp., Clostridium botulinum, Escherichia coli, Staphylococcus spp.*; virus such as Ebola virus, Hepatitis C virus, Tobacco mosaic virus, Human immunodeficiency virus; and fungi such as *Candida albicans, Puccinia striiformis, and Trichophyton rubrum*) possess great threat to plant, human, and animal health. Hence, there is considerable need of accurate and sensitive techniques to monitor these pathogens rapidly to maintain public health, prevent infection, and ensure compliance with standards and legislations. Conventional quantitative determination method involving culturing the pathogens in selective growth media has the major limitation of long-time consumption—such as *Listeria monocytogenes* may take up to 7 days, *Campylobacter fetus* may take more than 14 days to develop visible colonies. On the other hand, nucleic acid sequence-based assays (NASBA), such as PCR, DNA sequencing used for the confirmation of pathogens require purification step before identification, which increases the identification cost. Furthermore, nucleic acid amplification technique is useful to detect the presence of only known pathogens and not applicable to monitor toxin production. These drawbacks of conventional methods signify the need of development of rapid, sensitive, and accurate methods for pathogens and toxin detections (Byrne et al. 2009).

Bacterial cell detection using immunology-based methods prove to be powerful analytical tools. Immunomagnetic separation of targeted pathogens from the suspension of mixed culture conducted by magnetic beads coated with antibody is a preconcentration step, which is coupled with detection methods such as electrochemical, optical, magnetic, and magnetoresistance methods. ELISA is one of the common immunological method, which integrated sensitivity of enzymes and specificity of antibodies by coupling antibodies with the enzyme (Lazcka et al. 2007). For the effective determination of pathogens, antibody-based biosensor appears to be suitable analytical instrument because of its features such as (a) low detection limit—can detect bacteria in relatively small sample size (1–100 mL); (b) species selectivity—can distinguish between bacterial species; (c) strain selectivity—can distinguish between strains of same species; (d) low assay time—around 5–10 min for single test; (e) precision level (5%–7%); (f) no reagent requirement for assay; (g) direct measurement without the need of preenrichment; and (h) highly automated process (Ivnitski et al. 2000). Different types of antibody-based immunosensors utilized for different types of pathogen detection are summarized in Table 5.2.

TABLE 5.2

Examples of Antibody-Based Biosensor for Pathogen Detection

Biosensor	Pathogens	Detail	Reference
Bacteria			
Optical	*Lactobacillus spp* *Staphylococcus aureus*	Gold nanoparticles (AuNPs) were conjugated with anti-gram-positive bacteria antibody. The conjugate was used to detect gram-positive bacteria on the basis of AuNPs color change because of antibody–antigen interaction.	Verdoodt et al. (2017)
Electrochemical	*Staphylococcus aureus*	Single-walled carbon nanotubes conjugated with anti-*Staphylococcus aureus* antibodies were immobilized on electrode. The presence of *S. aureus* was detected on the basis of change in the peak current along with interaction of antigen and antibody.	Bhardwaj et al. (2017)
Surface plasmon resonance	*Salmonella enteritidis*	Antibody conjugated with magnetic nanoparticles was used for magnetic separation and detection purpose. Presence of *S. enteritidis* was detected on the basis of change in the refractive index associated with the binding of antigen to antibody.	Liu et al. (2016)
Virus			
Electrochemical	Hepatitis A virus	Hepatitis A virus was immobilized on carbon nano-powder paste and secondary antibody was labeled with peroxidase. Presence of virus was detected on the basis of amperometric current as analytical signal because of the addition of hydrogen peroxide and hydroquinone.	Mandli et al. (2017)
Magnetic	Avian influenza virus	Surface-enhanced Raman scattering (SERS)-based magnetic immunosensor was used to detect influenza virus in sandwich format.	Sun et al. (2017)
Fungal Pathogen			
Surface plasmon resonance	*Puccinia striiformis* spores	Mouse monoclonal antibody and surface plasmon resonance sensor were used for the detection of fungal spore.	Skottrup et al. (2007)
Voltammetric	*Phakopsora pachyrhizi*	Anti-mycelium of *P. pachyrhizi* antibody was immobilized on magnetic microbeads. Secondary antibody was labeled with phosphatase alkaline enzyme and pathogen detection was done on the basis of sandwich format.	

5.13 Antibody-Based Biosensor for Detection of Toxins

Immunoassays for the detection of bacterial toxins involve identification of target toxins by antibodies, followed by signal transduction and data processing. Toxins can be identified either by competitive or noncompetitive approach, based on the number of epitopes in toxins. In a competitive approach, there is a competition between immobilized antigens and free and labeled antigens to bind with limited number of antibody-combining sites such that the concentration of labeled antigen becomes inversely proportional to the concentration of free antigens with the progress of binding to the antibody. Low molecular weight toxins (monocyclic heptapeptide, microcystin) with one epitope are usually detected by competitive binding method. Noncompetitive assays can be conducted by two methods: (1) toxins are bound directly onto the transducer, which is quantified by using specific labeled antibodies and (2) two antibodies are used to bind and detect the toxin (sandwich immunoassays). After antibody-toxin binding, different methods can be used for signal transduction and readout generation. For toxin detection, different approaches such as SPR, electrochemical sensors, labeled immune reagents, and so on have been utilized along with fluorescence, luminescence, mass spectrometry, and electronic signal to improve the sensitivity of the process (Zhu et al. 2014).

5.13.1 Detection of Mycotoxins through Surface Plasmon Resonance

Most of the toxins that are detected with SPR have low molecular weight and indirect tests that offer better detection limits for their detection in samples. Several assays employing SPR for measuring mycotoxin concentration were described in Table 5.3.

Daly et al. (2000) used commercial SPR sensor Biacore 1000 to detect AFB1 on the basis of competitive inhibition immunoassay, where AFB1 was

TABLE 5.3

Detection of Some Mycotoxin through SPR Immunosensor

Toxin	MW(Da)	Type of Detection	Detection Limit
Deoxynivalenol	296.3	Indirect	0.5 ng/g
Deoxynivalenol	296.3	Indirect	2.5 ng/mL
Aflatoxin B1	312.3	Indirect	0.2 ng/g
Aflatoxin B1	312.3	Indirect	3 ng/g
Ochratoxin	403.8	indirect	0.1 ng/g
Ochratoxin	403.8	Direct	0.1 µg/mL
Fumonisin B1	721.8	Indirect	50 ng/g
Fumonisin	721.8	Direct	50 ng/mL

Source: Adapted from Hodnik, V. and Anderluh, G., *Sensors,* 9, 1339–1354, 2009.

conjugated to BSA and immobilized on dextran gel surface using amine coupling chemistry method. An organic solution consisting of ethanolamine and acetonitrile (pH 12.0) was used as an eluent and the mixture of the AFB1 and AFB1 polyclonal antibody was injected into the SPR sensor chip. The limit of quantification of the assay was 3 ng/mL, and the assay had a linear range of 3.0–98.0 ng/mL with good reproducibility.

Yu and Lai (2004) employed a polypyrrole (PPy) film on a miniaturized SPR device for the detection of Ochratoxin A (OTA). It is well known that PPy film, which is relatively easy to be prepared by the electro-oxidation and is stable, can absorb different biomolecules. In their experiment, PPy film doped with chloride (PPy–Cl) was electrochemically polymerized on SPR chip surface. By monitoring the SPR angle, the thickness of each PPy–Cl film could be controlled *in situ* from 2 to 5 nm. The PPy films were found to exhibit good binding capability with OTA in 10% methanol solution, and when they are used for the OTA detection based on an increase in SPR angle, the linear range extended from 0.1 to 10.1g/mL.

For the detection of OTA, SPR-immunosensor based on nano-size gold hollow ball (GHB) with dendritic surface has been developed. A thionine thin film was initially electropolymerized onto the SPR chip surface and then anti-OTA monoclonal antibody (anti-OTA) was immobilized onto the surface by GHB conjugation. The interaction of OTA and anti-OTA caused an increase in the resonant angle. The change of resonance angle is proportional to the concentration of OTA, and the LOD of OTA was 0.01 ng/mL (Fu 2007).

Yuan et al. (2009) detected OTA in cereals and beverages. Two different mixed self-assembled monolayers were compared through the effect of antibodies binding to OTA preimmobilized on SPR chip surface. One is the conjugate of OTA with BSA, and the other self-assembled monolayer is OTA–OVA conjugate with a PEG linker. The latter showed better binding capacity toward antibody than the former. A competitive inhibition immunoassay was carried out using the OTA–PEG–OVA surface. In addition, AuNPs (40 nm) were applied on the OTA–PEG–OVA surface for signal enhancement. The LOD can be significantly improved to 0.042 ng/mL for OTA, though OTA concentration of as low as 1.5 ng/mL could be directly detected on this surface. The surface exhibited high stability with more than 600 binding/regeneration cycles.

References

Alvarenga, L. M., J. de Moura, and P. Billiald. 2017. Recombinant antibodies: Trends for standardized immunological probes and drugs. In *Current Developments in Biotechnology and Bioengineering* (Ed.) V. Thomaz-Soccol, pp. 97–121. Tokyo, Japan: Elsevier.

Arruebo, M., M. Valladares, and Á. González-Fernández. 2009. Antibody-conjugated nanoparticles for biomedical applications. *Journal of Nanomaterials.* doi:10.1155/2009/439389

Bernareggi, D., S. Canevari, and M. Figini. 2014. Introductory remarks for the diagnostic and therapeutic applications of monoclonal antibodies and various formats. In *Autoantibodies* (Eds.) Y. Shoenfeld, P. L. Meroni, and M. E. Gershwin, pp. 83–90. Tokyo, Japan: Elsevier.

Bhardwaj, J., S. Devarakonda, S. Kumar, and J. Jang. 2017. Development of a paper-based electrochemical immunosensor using an antibody-single walled carbon nanotubes bio-conjugate modified electrode for label-free detection of food-borne pathogens. *Sensors and Actuators B: Chemical* 253: 115–123.

Božič, B., S. Čučnik, T. Kveder, and B. Rozman. 2014. Affinity and avidity of autoantibodies. In *Autoantibodies* (Eds.) Y. Shoenfeld, P. L. Meroni, and M. E. Gershwin, pp. 43–49. Tokyo, Japan: Elsevier.

Burlage, R. S. and J. Tillmann. 2017. Biosensors of bacterial cells. *Journal of Microbiological Methods* 138: 2–11.

Byrne, B., E. Stack, N. Gilmartin, and R. O'Kennedy. 2009. Antibody-based sensors: Principles, problems and potential for detection of pathogens and associated toxins. *Sensors (Basel, Switzerland)* 9: 4407–4445.

Crivianu-Gaita, V. and M. Thompson. 2016. Aptamers, antibody scFv, and antibody Fab' fragments: An overview and comparison of three of the most versatile biosensor biorecognition elements. *Biosensors and Bioelectronics* 85: 32–45.

Cruse, J. M. and R. E. Lewis. 1999. Antigen-antibody interaction. In *Atlas of Immunology* (Eds.) J. M. Cruse and R. E. Lewis, pp. 143–160. Berlin, Germany: Springer.

Daly, S. J., G. J. Keating, P. P. Dillon et al. 2000. Development of surface plasmon resonance-based immunoassay for aflatoxin B1. *Journal of Agricultural and Food Chemistry* 48: 5097–5104.

Day, M. J. 2015. Introduction to antigen and antibody assays. *Topics in Companion Animal Medicine* 30: 128–131.

Fu, X. H. 2007. Surface plasmon resonance immunoassay for ochratoxin A based on nanogold hollow balls with dendritic surface. *Analytical Letters* 40: 2641–2652.

Hodnik, V. and G. Anderluh. 2009. Toxin detection by surface plasmon resonance. *Sensors* 9: 1339–1354.

Ivnitski, D., I. Abdel-Hamid, P. Atanasov, E. Wilkins, and S. Stricker. 2000. Application of electrochemical biosensors for detection of food pathogenic bacteria. *Electroanalysis* 12: 317–325.

Jang, H. S., K. N. Park, C. D. Kang et al. 2009. Optical fiber SPR biosensor with sandwich assay for the detection of prostate specific antigen. *Optics Communications* 282: 2827–2830.

Jazayeri, M. H., H. Amani, A. A. Pourfatollah, H. Pazoki-Toroudi, and B. Sedighimoghaddam. 2016. Various methods of gold nanoparticles (gnps) conjugation to antibodies. *Sensing and Bio-Sensing Research* 9: 17–22.

Kennedy, P. J., C. Oliveira, P. L. Granja, and B. Sarmento. 2017. Antibodies and associates: Partners in targeted drug delivery. *Pharmacology and Therapeutics.* In Press.

Lazcka, O., F. J. D. Campo, and F. X. Muñoz. 2007. Pathogen detection: A perspective of traditional methods and biosensors. *Biosensors and Bioelectronics* 22: 1205–1217.

Li, Y., X. Liu, and Z. Lin. 2012. Recent developments and applications of surface plasmon resonance biosensors for the detection of mycotoxins in foodstuffs. *Food Chemistry* 132: 1549–1554.

Liu, X., Y. Hu, S. Zheng, Y. Liu, Z. He, and F. Luo. 2016. Surface plasmon resonance immunosensor for fast, highly sensitive, and in situ detection of the magnetic nanoparticles-enriched *Salmonella enteritidis*. *Sensors and Actuators B: Chemical* 230: 191–198.

Lyon, L. A., D. J. Pena, and M. J. Natan. 1999. Surface plasmon resonance of Au colloid-modified Au films: Particle size dependence. *The Journal of Physical Chemistry* 103: 5826–5831.

Ma, H. and R. O'Kennedy. 2017. Recombinant antibody fragment production. *Methods* 116: 23–33.

Mandli, J., A. Attar, M. M. Ennaji, and A. Amine. 2017. Indirect competitive electrochemical immunosensor for hepatitis A virus antigen detection. *Journal of Electroanalytical Chemistry* 799: 213–221.

Mazur, F., M. Bally, B. Städler, and R. Chandrawati. 2017. Liposomes and lipid bilayers in biosensors. *Advances in Colloid and Interface Science*. In Press.

Meulenberg, E. P. 2012. Immunochemical methods for ochratoxin A detection: A review. *Toxins (Basel)* 4: 244–266.

Moticka, E. J. 2016. *Historical Perspective on Evidence-Based Immunology*. Tokyo, Japan: Elsevier.

Moulds, J. M. 2009. Introduction to antibodies and complement. *Transfusion and Apheresis Science* 40: 185–188.

Ndoja, S. and H. Lima. 2017. Monoclonal antibodies. In *Current Developments in Biotechnology and Bioengineering* (Eds.) V. Thomaz-Soccol, A. Pandey, and R. R. Resende, pp. 71–95. Tokyo, Japan: Elsevier.

Patris, S., M. Vandeput, and J. M. Kauffmann. 2016. Antibodies as target for affinity biosensors. *TrAC Trends in Analytical Chemistry* 79: 239–246.

Reverberi, R. and L. Reverberi. 2007. Factors affecting the antigen-antibody reaction. *Blood Transfusion* 5: 227–240.

Roda, A., M. Mirasoli, B. Roda, F. Bonvicini, C. Colliva, and P. Reschiglian. 2012. Recent developments in rapid multiplexed bioanalytical methods for foodborne pathogenic bacteria detection. *Microchimica Acta* 178: 7–28.

Schroeder, H. W. and L. Cavacini. 2010. Structure and function of immunoglobulins. *The Journal of Allergy and Clinical Immunology* 125: S41–S52.

Shankaran, D. R., K. V. Gobi, and N. Miura. 2007. Recent advancements in surface plasmon resonance immunosensors for detection of small molecules of biomedical, food and environmental interest. *Sensors and Actuators B: Chemical* 121: 158–177.

Shanker, R., G. Singh, A. Jyoti, P. D. Dwivedi, and S. P. Singh. 2014. Nanotechnology and detection of microbial pathogens. In *Animal Biotechnology: Models in Discover and Translation* (Eds.) A. S. Verma and A. Singh, pp. 525–540. Tokyo, Japan: Academic Press.

Singh, A., S. Chaudhary, A. Agarwal, and A. S. Verma. 2013. Antibodies: Monoclonal and polyclonal. In *Animal Biotechnology: Models in Discovery and Translation* (Eds.) A. S. Verman and A. Singh, pp. 265–287. Tokyo, Japan: Academic Press.

Skottrup, P., S. Hearty, H. Frøkiær, P. Leonard, J. Hejgaard, R. O'Kennedy, M. Nicolaisen, and A. F. Justesen. 2007. Detection of fungal spores using a generic surface plasmon resonance immunoassay. *Biosensors and Bioelectronics* 22: 2724–2729.

Sonezaki, S., S. Yagi, E. Oqawa, and A. Kondo. 2009. Analysis of the interaction between monoclonal antibodies and human hemoglobin (native and cross-linked) using a surface plasmon resonance (SPR) biosensor. *Journal of Immunological Methods* 238: 99–106.

Stenberg, E. S. A., B. Persson, H. Roos, and C. Urbaniczky. 1991. Quantitative determination of surface concentration of protein with surface plasmon resonance using radiolabeled proteins. *Journal of Colloid and Interface Science* 143: 513–526.

Strohl, W. R. and L. M. Strohl. 2012. *Therapeutic Antibody Engineering*. New Delhi, India: Woodhead Publishing.

Sun, Y., L. Xu, F. Zhang, Z. Song, Y. Hu, Y. Ji, J. Shen, B. Li, H. Lu, and H. Yang. 2017. A promising magnetic sers immunosensor for sensitive detection of avian influenza virus. *Biosensors and Bioelectronics* 89: 906–912.

Taylor, A. D., J. Ladd, Q. Yu, S. Chen, J. Homola, and S. Jiang. 2006. Quantitative and simultaneous detection of four foodborne bacterial pathogens with a multi-channel SPR sensor. *Biosensors and Bioelectronics* 22: 752–758.

Velusamy, V., K. Arshak, O. Korostynska, K. Oliwa, and C. Adley. 2010. An overview of foodborne pathogen detection: In the perspective of biosensors. *Biotechnology Advances* 28: 232–254.

Verdoodt, N., C. R. Basso, B. F. Rossi, and V. A. Pedrosa. 2017. Development of a rapid and sensitive immunosensor for the detection of bacteria. *Food Chemistry* 221: 1792–1796.

Vo-Dinh, T. and B. Cullum. 2000. Biosensors and biochips: Advances in biological and medical diagnostics. *Fresenius' Journal of Analytical Chemistry* 366: 540–551.

Yu, J. C. C. and E. P. C. Lai. 2004. Polypyrrole film on miniaturized surface plasmon resonance sensor for ochratoxin A detection. *Synthetic Metals* 143: 253–258.

Yuan, J., D. Deng, D. R. Lauren, M. I. Aquilar, and Y. Wu. 2009. Surface plasmon resonance biosensor for the detection of ochratoxin A in cereals and beverages. *Analytica Chimica Acta* 656: 63–71.

Zhu, K., R. Dietrich, A. Didier, D. Doyscher, and E. Märtlbauer. 2014. Recent developments in antibody-based assays for the detection of bacterial toxins. *Toxins* 6: 1325–1348.

6

Bionanofabrication and Bionano Devices in Tissue Engineering and Cell Transplantation

6.1 Tissue Engineering

Different body tissues or organs are susceptible to damage by congenital disease, trauma, or accidents. The self-repair mechanism of the body can regenerate or repair the damaged tissues; however, the pace of the recovery is slower and is limited to only minor damage. In case of severe damage, more complicated processes, surgery, prosthesis, drug therapy, or transplantation are required (Gorain et al. 2017). However, full restoration of the damaged tissue is difficult and the result might not be functionally or esthetically satisfactory. Transplantation of organs or tissues from the donor to recipient is an emerging lifesaving technology and depending on donor and recipient organs, transplantations are classified as follows: (1) autotransplants, that is, transplantation of tissue within same patients from one site to other, (2) allotransplants, which involves transplanting tissue from one individual to another within same species, and (3) xenotransplants, which involves transfer across species barrier (Bakari et al. 2012). However, organ transplantation techniques have several constraints such as lack of the donor tissue, rejection of implanted organs by human immune system, disruption of body function, inflammation, tumor formation, and destruction over the period. The recipient immune system recognizes the transplanted tissue as foreign and imparts effector mechanisms that reject and ultimately destroy the implanted organs. On the basis of time period, rejection can be hyperacute, acute, or chronic rejection and based on the underlying mechanism of rejection, rejection can be either cell-mediated or antibody-mediated (Phillips and Callaghan 2017).

Tissue engineering is an interdisciplinary field that applies the principles of both engineering and the processes/phenomena of the life science such as mechanical engineering, clinical medicine, material science, and genetics to restore, maintain, and improve tissue functions. The new and functional tissues are fabricated in living cells such as stem cells, utilizing matrix or scaffolds for tissue culture. This emergence of stem cell biology has introduced a term known as *regenerative medicine* (Vacanti and Vacanti 2011). Tissue engineering represents a novel scientific approach for the regeneration of patients' own tissues from cells with the help of scaffolds, biomaterials, and growth factors. This approach counteracts the problems faced by the conventional donor organ transplantation such as need of donor organ transplants, poor biocompatibility and biofunctionality, and immune rejection. It utilizes three factors: (1) bioactive molecules such as growth factor to induce tissue growth, (2) cells that respond to various signals, and (3) seeding of cells into three-dimensional (3D) matrices to create tissue-like constructs to replace the lost parts of tissues or organs (Akter 2016).

Tissue engineering extensively uses porous 3D scaffolds to provide the appropriate environment for the regeneration of tissues and organs. These scaffolds seeded with cells and growth factors act as a template for tissue formation, which are cultured either *in vitro* to synthesize that can be implanted into an injured site, or *in vivo* and are implanted directly into the injured site where they are regenerated using the body's own systems (O'Brien 2011). The combination of cells, signals, and scaffold is regarded as tissue engineering triad as illustrated in Figure 6.1.

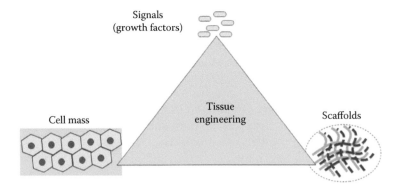

FIGURE 6.1
The basic paradigm of tissue engineering. (From Ivkovic, A. et al., *Front. Biosci.*, 1, 923–944, 2011. With Permission.)

6.2 Fabrication of Micro- and Nanodevices for Tissue Engineering

Tissue engineering, therefore, integrates biocompatible scaffolds, cells, active molecules, and drugs by the combined knowledge of cell biology, engineering, nanotechnology, and micro- and nanofabrication disciplines. In tissue engineering, nanomaterials are increasingly being important for the development of nanostructured scaffolds such as nanoparticles, nanotubes, nanofibers, and other fabricated nanostructured devices to mimic the native biological system for cell growth and tissue regeneration. Nanotechnology together with microfabrication and postprocessing modification approach develop wide range of two-dimensional and 3D biocompatible nanostructured material to meet the requirements of native tissue recovery for *in vitro* tissue engineering and implantation application (Limongi et al. 2016).

Micro- and nanoscale technologies are emerging tools for tissue engineering and are used to fabricate biomimetic scaffolds with higher vascularization and complexity and to control the cellular microenvironment such as cell-to-cell, cell-to–matrix, and cell-to-soluble factor interactions (Chung et al. 2007). Recently, tissue engineering is utilizing nanotechnology for cell tracking via nano-based labeling techniques for noninvasive *in vivo* imaging. For this purpose, two types of nanoparticles, that is, magnetic nanoparticles and fluorescent nanoparticles are being utilized. Magnetic nanoparticles consist of superparamagnetic iron oxide (SPIO) particles with 50–500 nm in diameter and fluorescent cell-labeling techniques using Quantum dots nanoparticles with 2–5 nm diameters have been applied for cell tracking. Electrospinning and self-assembly are the commonly utilized method for fabrication of nanofibrous biomaterials (Bean and Tuan 2013).

6.3 Scaffold Fabrication Methods

The fabrication of scaffold should allow the control over porosity, pore size, shape, and size but should not affect the biocompatibility of material or alter the physical, chemical, and biological properties of the materials. Various fabrication methods ranging from traditional methods such as fiber bonding, solvent casting, and advanced computer-based techniques such as 3D printing have been used. The common scaffold fabrication methods are summarized in Table 6.1.

TABLE 6.1

Different Scaffold Fabrication Methods

S. No.	Fabrication Process	Details
1	Fiber bonding	Fibers are bonded at cross points using secondary polymer resulting in 3D porous structure.
2	Solvent casting/ Particulate leaching	Polymers along with porogens are dissolvent in organic solvent. When solvent evaporates polymer, matrix is formed from which salt leaches out the resulting porous structure.
3	Melt molding	Polymers are heated above glass-transition temperature, thus can be molded into shapes
4	Membrane lamination	Polymeric membranes are formed by stacking membranes.
5	Phase separation	Polymer-rich phase and poor phase are separated by liquid–liquid phase separation. Polymer-rich phase solidifies and polymer-poor phase is removed forming a porous polymer network.
6	Freeze drying	Polymer solvent solution is homogenized to form an emulsion, which is rapidly cooled to lock liquid-state structure. Solvent is removed by freeze-drying.
7	Gas foaming	Polymer foam is formed by exposing polymers to foaming agents at high pressures. Reduction in pressure causes removal of foaming agents resulting in nucleation and growth of pores in polymer.
8	Polymer ceramic composite foam	Ceramics are incorporated into poly (α-hydroxy ester) polymer by solvent-casting method.
9	Solid-free form techniques	Material in powder or liquid form is solidified in layers by a computer program such as laser sintering, stereolithography, 3D printing, fused deposition modeling.
10	Selective laser sintering	Solvent-free approach where laser beam is used to heat polymeric powder causing fusion of polymeric particles.
11	Fused deposition modeling	3D scaffold is deposited layer-by-layer through a nozzle attached to a computer-connected device.
12	3D printing	Adhesive solution is deposited on polymeric powder bed by inkjet printer forming porous scaffolds.
13	Pressure-assisted microsyringe	Polymer is dissolved in solvent and deposited through a syringe fitted with a 10–20 µm glass capillary needle.

Source: Adapted from Verma, P. and Verma, V., Concepts of tissue engineering, in *Animal Biotechnology*, Verma, A.S. and Singh, A. (Eds.), Academic Press, Tokyo, Japan, pp. 233–245, 2014.

6.4 Microfabrication and Nanolithography

Wide range of microfabrication methods has been developed to create efficient biomimetic devices in tissue engineering. Microfabrication approaches such as micromachining, photolithography, metal deposition, electrospinning, wet and dry etching, thin-film growth, and 3D printing were used to design the features of different materials and surfaces on the micron and submicron scale (McMahon et al. 2011), whereas techniques such as electron beam and focused ion-beam lithography were used to design the structures in nanometer range (Raffa et al. 2008).

6.4.1 Replica Molding

Microfabrication methods such as optical lithography, deposition, and etching techniques designed for silicon and glass are not suitable alone to fabricate natural and synthetic biomaterials; thus, they need to be coupled with other technologies. Replica molding and embossing approach are being used to fabricate nanoscale structures. Nanoscale structures are usually optimized by high-resolution master, conventional nanofabrication approach, which can be replicated by molding or embossing. Replica molding and soft lithography technology can develop nanometric replicas of biocompatible polymers such as polydimethylsiloxane, polystyrene, and so on. Hot embossing, solvent casting, thermoforming, and injection molding can generate nanoscale thermoplastic biocompatible materials (Limongi et al. 2016). Electroforming of nickel against nanoporous anodic aluminum oxide followed by nanoinjection molding or hot embossing can be applied in neural differentiation (Jung et al. 2015).

6.4.2 Etching and Direct-Write Method

Etching away bulk materials is a top–down fabrication approach, which results in nanostructured materials with precise morphology, size, and functional groups. Micro/nanoscale-shaped structures with continuous, self-aligned, fiber–fiber and template-free manner fibers have been fabricated by electrospinning and novel direct-write 3D electrospinning are increasingly showing potential in tissue engineering application. Plasma etching represents nanolithography techniques, which in combination with soft, nanoimprint, and dip-pen lithography and conventional approaches such as Electron beam lithography (EBL) and Focused ion beam lithography (FIBL) have applications for nanofabrication of biomaterials (Limongi et al. 2016).

Plasma etching is a surface treatment method, which can modify the surface properties of biomaterials and enhance the biocompatibility without affecting the bulk properties. This fabrication method is becoming popular in tissue engineering through the modification of the surfaces of implantable devices and components. For instance, nanoscale PCL films fabricated in silicon wafer via single-step plasma etching remained functioning in culture media for weeks, and showed adhesive properties as well (Cesca 2014).

6.5 Strategies of Tissue Engineering

Three strategies, that is, cell injection, cell induction, and cell-seeded scaffold are described as the approaches of tissue engineering. All these approaches depend on key elements; for example, cells, growth factors, and matrix to guide tissue regeneration.

6.5.1 Cell Injection

Cell injection therapy relies on the principle of tissue formation from cellular action and implies injection of stem cells into the defect to regenerate tissues. However, the effectiveness of this approach is limited by low engraftment, inadequate localization of injected cells because of continuous movement, immunological rejection, and the maintenance of phenotype by injected cells. Live cells encapsulated into a delivery vehicle are being used for the prevention of direct contact with the immune system and thus enhance adequate localization (Ravichandran et al. 2012) for effective therapy.

6.5.2 Cell-Induction Therapy

Cell-induction therapy is another approach for tissue engineering. Osteoinduction therapy used for the regeneration of craniofacial bone required no exogenous biological components other than signaling molecules; growth/differentiation factors such as fibroblasts growth factors—2 and 9, transforming growth factors, vascular endothelial growth factors, recombinant human growth factors, and bone morphogenetic protein are used to induce osteogenesis (Miron and Zhang 2012).

6.5.3 Cell-Seeded Scaffold

Cell-seeded scaffold strategy depends on the isolation of appropriate cell population from patients or donor, usually mesenchymal stem cell (MSC), which has the potential to differentiate into tissue-specific cells for regenerative medicine and also possesses important immunomodulatory properties for treating immune-related diseases. MSC can regulate the intensity of immune response by inducing T-cell apoptosis and therefore can have great therapeutic potential for tissue engineering. The cells are seeded within or onto a natural or synthetic scaffold where cells will proliferate, differentiate, and regenerate into *organoid*, which are then implanted into the patient. In another strategy, acellular scaffolds are implanted into the defect while the body cells can populate the scaffold to form the new tissue *in situ* (Aou Neel et al. 2014).

6.6 Principles of Tissue Engineering

The general approach of tissue engineering is isolation of cells from target tissue, preparation of single-cell suspension, followed by growth of cells onto a 3D synthetic ECM, scaffold with appropriate growth factors permitting the formation of 3D mass with specific biochemical or phenotypic properties of tissues. In natural condition, cells are organized into 3D forms surrounded by natural extracellular matrix (ECM) such as collagen, elastin, proteoglycans, glycosaminoglycans, glycoproteins, hyaluronic acid, and so on, which helps to maintain spaces between cells, bind the cells in tissues, and act as reservoir of hormones, growth factors, enzymes, and others. Tissue engineering focuses on creating artificial ECM around the cells of specific tissues. Cells, scaffolds, media, and bioreactors are major requirements of tissue engineering (Post and Weele 2014).

6.6.1 Cell

The success of tissue engineering relies largely on the source of the cells and depending on the source, the cells are classified as autologous, allogeneic, and xenogeneic. The autologous cells are obtained from the same individual to be implanted, allogeneic cells are obtained from same species other than the recipient, and xenogeneic cells are obtained from individual of different species because of which allogeneic and xenogeneic cells have the risk of host rejection, transmission of disease, and associated ethical issues (Verma and Verma 2014).

Allogeneic tissues have been extensively used for immunosuppressive drug therapies, connective tissue replacement, and cartilage and skin tissue engineering. Xenotransplantation using xenogeneic cells have been used in different clinical specialties such as injection of porcine islet cells of Langerhans into patients with type 1 diabetes mellitus, and transplantation of pig neuronal cells into patients with Parkinson's disease and Huntington's diseases. Autologous cell therapy implies harvesting autologous cells from an individual followed by culturing and reintroducing into the damage site (Akter 2016). However, this therapy has numerous limitations such as risk of cell being at a diseased state and invasive nature of collection, because of which stem cells such as embryonic stem cells (ESCs), MSCs, bone marrow mesenchymal stem cells (BMMSCs), and umbilical cord derived mesenchymal stem cells (UCMSCs) have gained much attention (Howard et al. 2008).

A stem cell is pluripotent in nature, which can proliferate into any cell type and have unlimited ability to renew itself. Stem cells can be either unipotent, giving rise to single cell type or multipotent, which can give rise to all types of cells. According to source, stem cells are categorized as ESC, adult stem cells, induced pluripotent stem cells, neural stem cells, hematopoietic stem cells, and skin stem cells. Pluripotent ESCs can generate all tissues in an organism, multipotent neural stem cells exhibit self-renewal properties and the ability to differentiate into all neural subtypes. Induced pluripotent stem cells are the cells reverted to stem cell. ESCs have two major properties: (1) ability to proliferate in self-renewal (pluripotent) state but remains undifferentiated and (2) ability to differentiate into many specialized cells, that is, induced pluripotent. Stem cells obtained from somatic cells can be used to design patient-specific cell therapies, adult stem cells have the ability to differentiate into many different types of tissues, such as osteoblasts, chondrocytes, adipocytes, myocytes, and tenocytes (Olson et al. 2011).

Other types of cells that are utilized include immortalized cells and primary cells. Immortalized cells are the cells that are tumorous or are induced to proliferate indefinitely and are utilized for cell culture. However, the major drawbacks of using immortalized cells are that these cells are not normal, and therefore sometimes may express gene patterns that are not present in natural conditions and the cells are prone to change after several passages. Primary cells are obtained directly from tissue removed from living animals; thus, they allow direct investigation of cell under controlled *in vitro* conditions. However, primary cells have disadvantages of limited life span and heterogeneous population with high variability (Carter and Shieh 2015).

6.6.2 Scaffolds

Scaffolds are the artificial ECMs and similar to natural ECMs, they help in proliferation, differentiation, and cell synthesis. Hence, scaffolds direct the cells to maintain a 3D organization and support the formation of engineered tissues. Different scaffolds have been developed from biomaterials using a plethora of fabrication methods to assist in regeneration of tissues and organs. For designing suitable scaffold, numerous key factors need to be considered (Sensharma et al. 2017). Some of the ideal properties of scaffolds to be considered are summarized in Table 6.2.

TABLE 6.2

Ideal Characteristics of Scaffolds

Properties	Details
Biocompatibility	Scaffold must be biocompatible allowing cell adherence and proliferation onto surfaces. Scaffold should not exhibit immune reaction, which may cause rejection or healing reduction.
Biodegradability	Scaffold must be biodegradable to permit production of natural ECM. Resulting by-product from degradation by macrophage must be nontoxic and exit the body without harming other body organs.
Mechanical properties	Scaffold must be strong to withstand surgical handling. Scaffold must exhibit sufficient integrity from the time of implantation to the completion of the remodeling process.
Scaffold architecture	Scaffold must have interconnected pore structure and high porosity to induce cellular penetration and diffusion of nutrient, waste products, and degradation product of scaffolds. Scaffold must have a critical pore range depending on the cell type and tissue being engineered.
Manufacturing technology	The manufacturing cost should be minimum and should be easy to scale-up

Source: Adapted from O'Brien, F.J., *Mat. Today.*, 14, 88–95, 2011.

6.6.3 Biomaterials

Biomaterials are the materials intended to interface with biological systems for evaluating and replacements of tissues, organs, or function of the body. Biomaterial serves as artificial ECM and replicate the biological and mechanical function of the native ECM. It provides 3D space for cell proliferation, cell adhesion, substrate to delivery of cells, and bioactive factors to the specific site of action and mechanical support against *in vivo* forces to predefined 3D structure and bioactive signal, including adhesive proteins and growth factors (Rustad et al. 2010).

Ideal biomaterials that are used to fabricate scaffolds should be biodegradable, bioresorbable, and should have the ability to provide conditions for the regulation of cell behavior. Three types of biomaterials: (1) natural biomaterials such as chitosan, alginate, collagen, and proteoglycans; (2) synthetic biodegradable polymers such as polylactic acid (PLA), polyglycolic acid (PGA), and poly-dl-lactic-co-glycolic acid (PLGA); and acellular tissue matrices such as submucosa and small intestinal submucosa have been used to fabricate scaffolds for tissue engineering. Natural and decellularized matrices have the advantages of biological recognition, whereas synthetic polymers have the advantages of large-scale production with controlled properties of strength, degradation rate, and microstructure (Lee and Atala 2014).

Ceramic scaffolds such as hydroxyapatite (HA) and tricalcium phosphate (TCP) have high mechanical stiffness, low elasticity, and brittle surface and are used for making porous scaffolds for bone tissue regeneration, as they are known to enhance osteoblast differentiation and proliferation. Nano- and bioglass-based ceramics are also being used for this purpose. Recently, organic conductive biomaterials have revolutionized biomedical field because of great electrical and magnetic properties. The conductive polymers such as polyaniline, polypyrrole, and poly(3,4-ethylenedioxythiophene) are found to be better biomaterials over the graphite-based inorganic materials (Gajendiran et al. 2017).

6.6.4 Growth Media

Media aim to supplement nutrition to the cells and complement scaffolds to complete bioartificial ECM. On the basis of the function, media are clas- sified as seeding media, differentiation media, and maintenance media. Seeding media help to introduce cells into the scaffold, differentiation media containing growth factors, cytokines, help to differentiate cells into tissue, and maintenance media containing serum help to support cells in culture (Verma and Verma 2014).

Growth factors such as bone morphogenetic proteins, basic fibroblast growth factor, vascular epithelial growth factor, and transforming growth factor are the signaling molecules that control cell responses by binding of transmembrane receptors on target cells and promote in tissue regeneration. Usually growth factors are immobilized to scaffold by either noncovalent or covalent binding to prevent loss of bioactivity of growth factors (Akter 2016).

Important factors to be considered during cell culture are composition of culture media; temperature, oxygen, and carbon dioxide concentration; pH and osmolality. Therefore, the important step in cell culture is the selection of appropriate culture media, which is usually composed of basic compo- nents: amino acids (nitrogen source); vitamins; inorganic salts (Ca^{+2}, Mg^{+2}, Na^+, K^+); glucose and fructose (energy source); fat and fat-soluble compo- nents (fatty acids, cholesterols); nucleic acid precursors; serum as source of growth factors and hormones; antibiotics; pH and buffering system; oxygen and carbon dioxide concentration; and attachment factors. For mammalian cell culture, culture media can be either natural media composed of coagu- lant, tissue extracts, and biological fluids such as plasma, serum, or synthetic media composed of organic and inorganic nutrients, vitamins, salts, serum proteins, carbohydrates, and cofactors. Artificial media are further classified into four groups: (1) serum-containing media, (2) serum-free media, (3) chem- ically defined media, and (4) protein-free media. Serum-containing media is a complex mixture of small and large molecules, amino acids, growth fac- tors, vitamins, proteins, hormones, lipids, and minerals. Serum-free media are simple with well-defined composition depending on the cell type result- ing in easier downstream processing of products. Chemically defined media contain pure inorganic and organic constituents together with other pro- teins, vitamins, fatty acids, and growth factors. Protein-free media contain nonprotein constituents, which facilitates superior cell growth and efficient downstream purification of expressed products (Verma 2014).

Mammalian cells have been cultured in different basal media such as Eagle's minimum essential medium, Dulbecco's modified MEM, 199 media, and so on. Serum-free or chemically defined media are more useful than serum containing media to produce protein in very small amounts. The role of serum was hypothesized to supply hormones and growth factors rather than nutrients. Therefore, serum-free media containing peptide hormones and different growth factors such as platelet growth factor, nerve growth factor, transforming growth factor, and others have been designed (Sato 2014).

Different culture media with variation based on pH, nutrient concentration, and presence or absence of growth factor are available to supply nutrients and energy for cells during cell culture. Culture media should maintain isotonic condition between cell and media and are buffered at a pH, usually 7.4. Serum may be added to culture media to promote survival of the cells in presence of undefined mixture of growth factors, hormones, and proteins, such as platelet-derived growth factor, insulin, and transferrin. Serum-free media such as N2 or B27 containing known formulation of growth factors can be used to practice strict control over the cell culture. Certain growth factors such as epidermal growth factors and fibroblast growth factors are added in culture media to maintain neural progenitor pool and prevent differentiation (Carter and Shieh 2015).

6.6.5 Bioreactors

Bioreactors are the devices that use mechanical means to influence biological processes under controlled operating conditions. Bioreactors help to maintain uniform distribution of cells in the scaffold, maintain nutrition, and dissolved oxygen concentration. They are usually used to improve cell-culture efficiency by providing signals for growth and differentiation and reduce the operation time (Verma and Verma 2014).

6.7 Cell Culture

The term *cell culture* refers to the process of growing and maintaining cells outside living animals under controlled condition. This *in vitro* approach is desirable to simplify the cellular environment, practice control over experimental manipulation, and reduce interaction with other biological processes. *In vitro* techniques are preferred to *in vivo* because the experiment that is difficult to perform in an intact organism is simplified, and the process is relatively faster and cheaper requiring lesser number of animals (Carter and Shieh 2015).

6.7.1 Primary Cell Culture

To cultivate the cell culture, primary cells, cells isolated directly from a body by biopsy, surgery, or autopsy, are cultured for a finite time to produce primary cell cultures, which might be slightly different from primary cells. Primary cell cultures are obtained via explant culture, which involves the removal of ECM by mechanical means such as mincing or by chemical means such as enzyme digestion followed by the incubation of tissue on growth surface along with growth medium. Under light microscopic examination, if cells are observed in an area surrounding tissue mounds, mounds are removed and remaining cells are allowed to proliferate and the cell culture obtained is known as primary cell cultures (Godbey 2014).

Different morphological structures of cells in culture include the following: (1) epithelium type, which are polygonal shape and forms monolayer on solid substrate; (2) epithelioid cells, which are round and do not attach and form layer onto surface; (3) fibroblast type, which are angular shaped and form network of cells; and (4) connective tissue type, which contain large amount of fibrous and amorphous substances derived from fibrous tissue. Primary cell culture represents the best experimental models for *in vivo* analysis and share same karyotypes as parent; however, they are difficult to isolate, have limited life span, and have high risk of contamination by viruses and bacterial cells (Verma 2014).

6.7.2 Adherent Culture and Nonadherent Culture

In adherent culture, cells remain attached to the substrate via adhesion proteins and integrins. Proteins such as fibronectin, vitronectin, osteopontin, collagens, thrombospondin, fibrinogen, and so on, can be used to enhance the cell adherence (Godbey 2014). Mouse fibroblast STO cells are adherent cell types, which are difficult to grow as cell suspension. In case of suspension culture, cells do not need surface for an attachment but are freely floating in the medium. The source of cells is the governing factor for suspension cells. Blood cells suspended in plasma can be easily grown in suspension cultures (Verma 2014).

6.7.3 Secondary Cell Culture

When the primary cell culture expands, it may be passaged or subcultured into fresh culture media and the new culture is known as secondary culture. The passaging or subculturing is done by enzymatic digestion followed by washing and resuspending of cells in fresh growth media. Secondary cell culture has a longer lifespan than primary cell culture because of the availability of nutrient at regular intervals. They are also easy to grow and useful in virological, immunological, and toxicological studies. It is suitable to develop a large population of similar cells; however, the cells have a tendency to differentiate over a period of time and develop aberrant cells (Verma 2014).

6.8 Tissue Culture

The term *tissue culture*, coined by the American pathologist Montrose Thomas Burrows, is defined as the growth of tissues outside the living organism supported by the growth medium. The term *tissue culture* commonly refers to the culture of animal cells and tissue, whereas *plant tissue culture* is used for plant cells and tissues (Carrel and Burrows 1910).

6.8.1 Animal Tissue Culture

In biomedical sector, animal and human tissue cultures have wide range of scope ranging from the production of biopharmaceutical products, monoclonal antibodies, and gene therapy product. It also serves as a system to study biochemical pathways, intra- and intercellular responses, pathological mechanisms, and virus production. The progress in DNA technology and tissue culture has led to the development of recombinant vaccine against hepatitis B virus. In viral production, two stages are involved: (1) cell culture system, which leads to the development of efficient system to convert culture medium into cell mass and (2) virus production utilizing numerous immortalized cell lines. Production of virus-like particles is one of the new approaches of animal tissue culture. Recombinant therapeutic proteins such as cytokines, hematopoietic growth factors, growth factors, hormones, blood, enzymes, and antibodies can be assisted by tissue culture. In case of gene therapy, initially, a faulty gene is identified followed by the isolation and correction outside patient's body in the cultured cells. The corrected genes are then expanded and transferred back to the individual (Verma 2014).

6.8.2 Plant Tissue Culture

Plant tissue culture is defined as "the *in vitro* culture of plant protoplasts, cells, tissues, or organs under controlled aseptic conditions, which lead to cell multiplication or regeneration of organs whole parts." Plant tissue culture expresses the totipotency and induces genotypic and phenotypic manipulation in plant cells (Bhatia 2015a).

Plant tissue culture comprises excising plant tissues and growing them on nutrient media. It is a broad term that includes different approaches such as meristem culture for propagation of virus-free plants, protoplast culture, cell suspension culture, tissue and organ culture, and production of haploid culture by anther or pollen culture. Plant tissue culture improves the crop attributes by the general approach: initiation and establishment of culture, followed by techniques such as clonal propagation, virus elimination, selection of variants, genetic transformation, and finally regeneration of selected plants (Kumar and Loh 2012).

Plant cells have the ability to dedifferentiate, proliferate, and regenerate into mature plants. The idea of *in vitro* growth of plant cells was pioneered by Haberlandt and the concept has enormous application in basic and applied research. In earlier days, plant tissue culture was focused on basic research such as study of cell division, plant growth, and biochemistry. With the advancement of genetic engineering and technology, the scope of tissue engineering has widened to molecular pathways, tissue morphology, biochemical, and somatic cell genetics. And the approaches such as transgenic, somatic hybridization, mutant selection, somatic embryogenesis, metabolic engineering, organogenesis, molecular farming, and phytoremediation have enhanced the application of *in vitro* technology in plant science. The application of plant tissue culture is broadly divided into three categories: (1) basic research, (2) environmental issues, and (3) commercial application. Basic research area includes the study of physiology and molecular pathways in plant cells, environmental application includes techniques to preserve the elite germplasms for a longer time, and commercial application involves crop improvement, production of secondary metabolites, and genetic improvement (Bhatia 2015b).

6.9 Three-Dimensional Printing

The concept of 3D printing was first introduced in 1986 by Charles W. Hull and has been referred as additive manufacturing or rapid prototyping. 3D printing is one of the additive manufacturing processes and is used to make by adding layer-by-layer (LbL) nanodeposition of materials. This is associated with 3D modeling software computer-aided design (CAD) or computer tomography (CT) scan image, machine equipment, and layering materials. After CAD sketch, printing equipment read CAD file and 3D structure is created (Gu et al. 2016).

Initially, the appearance and proportions of the objects designed in a CAD program (Autodesk, AutoCAD, and Solidworks) are converted to standard Tessellation Language (STL) or Stereolithography file format, which are further converted to G-files and are subsequently transmitted to the printing machine. G-file functions to divide STL file into the series of two-dimensional horizontal cross sections resulting in thin layers, which are further developed into 3D structure. In general, 3D printer initially follows the instruction in the CAD file to make the foundation along the x–y plane. The printer then continues to move the print-head along the z-axis to build the object vertically LbL (Ventola 2014).

6.10 3D Printing in Tissue Engineering

On account of recent advancement in 3D printing technique, it is being used for the development of transformative tool for biomedical applications in the field of tissue engineering and regenerative medicine. Tissue engineering aims to produce biological substitutes of native tissue, *in vitro* drug screening to reduce animal specimen, *in vivo* transplantation to overcome organ shortage, and transplantation need. A 3D cell culture produces relevant physiological conditions for cell differentiation and proliferation, which explain the need of 3D biofabrication techniques to develop native biological components and growth conditions.

Traditional scaffold fabrication approaches including electrospinning, freeze-drying, gas foaming, and particle leaching have control on bulk properties but have the major limitation of lack of control on the architecture and topology of the scaffold. In photolithography techniques, photomasks used for polymerization of photosensitive biomaterial to develop 3D cell-laden hydrogel scaffolds have major drawbacks of need of multiple photomasks and different molds to develop different design 3D structures. However, 3D printing assisted with CAD techniques does not require physical masks and mold to develop internal architecture, thus this approach has major role as a cheaper and effective transformative tool for obtaining 3D structures at micro/nanoscale (Zhu et al. 2016).

Inkjet bioprinting, extrusion bioprinting, and light-assisted bioprinting are major approach for 3D printing of biomaterial and cells. Inkjet bioprinters print either individual cells or small clusters. This technology has the advantages of being cheap, rapid and versatile but suffers with the disadvantage of assurance of high cell density needed for scaffold fabrication (Jakab et al. 2010). Inject bioprinter works by forming picolitre droplets and on the basis of the energy used for droplets formation, inkjet bioprinters are of two types: (1) thermal inkjet printers and (2) acoustic/piezoelectric bioprinters. The former one uses electric heat in the print head to produce air pressure for droplet formation and ejection, whereas the later used pulses formed by piezoelectric or ultrasound pressure for droplet ejection. Extrusion printer generates the pressure either by pneumatic or piston to extrude biomaterial beads through a micronozzle. Light/laser-assisted printer utilizes laser energy to generate pressure for the ejection of biomaterials containing cells onto a collector substrate. All types of printers precisely control x–y–z axes to fabricate the desired 3D biostructure. The schematic representation of 3D bioprinters is illustrated in Figure 6.2.

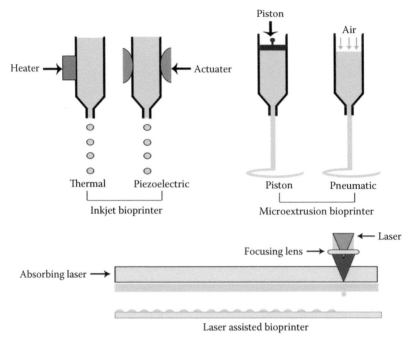

FIGURE 6.2
Schematic representation of different types of 3D printers. (From Zhang, X. and Zhang, Y., *Cell Biochem. Biophys.* 72, 777–782, 2015. With Permission.)

6.11 Current and Future Prospect of 3D Printing in Tissue Engineering

The rapid-prototyping approach, 3D bioprinting with application in industrial-scale fabrication, has a wide potential to overcome major limitation of conventional tissue-engineering approach. This technique can play important role in improving cellular integration, host integration, vascularization of 3D printed scaffolds, reduce steps, and purification cost to get the final product. Additional 3D printed tissues have scope in drug delivery and reduce the risk of pathological transition from animals to human. 3D printing approach is, therefore, a powerful tool to assemble functional tissue *in vitro* and accelerate the translation of tissue engineering in the development of regenerative therapy and drug testing (Jia et al. 2014).

Although 3D bioprinting techniques have numerous advantages, they need to overcome many challenges. Different research work, for instance, bioprinted human liver models with collagens, production of 3D printed scaffolds for human kidney, 3D printed ears using 3D bioprinting using collagen gels, and living cells have been developed by 3D bioprinting. Patient-customized 3D bioprinting is under development process. Further, in future development of biocompatible, easily available bioinks or biomaterials to be used for bioprinting, with efficient mechanical strength to support cell and secure 3D structure is the important aspect of 3D bioprinting. At present, bioinks are limited to gelatin, collagen, fibrin, ceramics, thermoplastics, or curable composite, which are used to prosthetic limbs, orthodontic devices, and bone implants printing. 3D printing has the future scope to be able to print soft tissue in veins and arteries and complete organs, which can be implanted directly into the human body (Gu et al. 2016).

References

Akter, F. 2016. *Tissue Engineering Made Easy*. Tokyo, Japan: Academic Press.

Aou Neel, E. A., W. Chrzanowski, V. M. Salih, H.-W. Kim, and J. C. Knowles. 2014. Tissue engineering in dentistry. *Journal of Dentistry* 42:915–928.

Bakari, A. A., U. S. A. Jimeta, M. A. Abubakar, S. U. Alhassan, and E. A. Nwankwo. 2012. Organs transplantation: Legal, ethical and Islamic perspective in Nigeria. *Nigerian Journal of Surgery* 18:53–60.

Bean, A. C. and R. S. Tuan. 2013. Stem cells and nanotechnology in tissue engineering and regenerative medicine. In *Micro and Nanotechnologies in Engineering Stem Cells and Tissues* (Eds.) M. Ramalingam, E. Jabbari, S. Ramakrishna, and A. Khademhosseini. Hoboken NJ: John Wiley & Sons.

Bhatia, S. 2015a. Plant tissue culture. In *Modern Applications of Plant Biotechnology in Pharmaceutical Sciences* (Eds.) S. Bhatia, K. Sharma, R. Dahiya, and T. Bera, pp. 31–107. Tokyo, Japan: Academic Press.

Bhatia, S. 2015b. Application of plant biotechnology. In *Modern Applications of Plant Biotechnology in Pharmaceutical Sciences* (Eds.) S. Bhatia, K. Sharma, R. Dahiya, and T. Bera, pp. 157–207. Tokyo, Japan: Academic Press.

Carrel, A. and M. O. N. T. R. O. S. E. Burrows. 1910. Cultivation of adult tissues and organs outside of the body. *Journal of the American Medical Association* 55:1379–1381.

Carter, M. and J. Shieh. 2015. *Guide to Research Techniques in Neuroscience*. Tokyo, Japan: Academic Press.

Cesca, F., T. Limong, A. Accardo et al. 2014. Fabrication of biocompatible free-standing nanopatterned films for primary neuronal cultures. *Royal Society of Chemistry* 4:4596–4702.

Chung, B. G., L. Kang, and A. Khademhosseini. 2007. Micro- and nanoscale technologies for tissue engineering and drug discovery applications. *Expert Opinion on Drug Discovery* 2:1653–1668.

Gajendiran, M., J. Choi, S. J. Kim et al. 2017. Conductive biomaterials for tissue engineering applications. *Journal of Industrial and Engineering Chemistry* 51:12–26.

Godbey, W. T. 2014. *An Introduction to Biotechnology*. Tokyo, Japan: Academic Press.

Gorain, B., M. Tekade, P. Kesharwani, A. K. Iyer, K. Kalia, and R. K. Tekade. 2017. The use of nanoscaffolds and dendrimers in tissue engineering. *Drug Discovery Today* 22:652–664.

Gu, B. K., D. J. Choi, S. J. Park, M. S. Kim, C. M. Kang, and C. H. Kim. 2016. 3-dimensional bioprinting for tissue engineering applications. *Biomaterials Research* 20:12.

Howard, D., L. D. Buttery, K. M. Shakesheff, and S. J. Roberts. 2008. Tissue engineering: Strategies, stem cells and scaffolds. *Journal of Anatomy* 213:66–72.

Ivkovic, A., I. Marijanovic, D. Hudetz, R. M. Porter, M. Pecina, and C. H. Evans. 2011. Regenerative medicine and tissue engineering in orthopaedic surgery. *Frontiers in Bioscience (Elite edition)* 1:923–944.

Jakab, K., C. Norotte, F. Marga, K. Murphy, G. Vunjak-Novakovic, and G. Forgacs. 2010. Tissue engineering by self-assembly and bio-printing of living cells. *Biofabrication* 2:022001.

Jia, J., D. J. Richards, S. Pollard et al. 2014. Engineering alginate as bioink for bioprinting. *Acta Biomaterialia* 10:4323–4331.

Jung, A. R., R. Y. Kim, H. W. Kim et al. 2015. Nanoengineered polystyrene surfaces with nanopore array pattern alters cytoskeleton organization and enhances induction of neural differentiation of human adipose-derived stem cells. *Tissue Engineering* 21:2115–2124.

Kumar, P. P. and C. S. Loh. 2012. Plant tissue culture for biotechnology. In *Plant Biotechnology and Agriculture* (Eds.) A. Altman and P. M. Hasegawa, pp. 131–138. Tokyo, Japan: Academic Press.

Lee, S. J. and A. Atala. 2014. Engineering of tissues and organs. In *Tissue Engineering Using Ceramics and Polymers* (Eds.) A. R. Boccaccini and P. X. Ma, pp. 347–348. Tokyo, Japan: Woodhead Publishing.

Limongi, T., L. Tirinato, F. Pagliari et al. 2016. Fabrication and applications of micro/nanostructured devices for tissue engineering. *Nano-Micro Letters* 9:1.

McMahon, R. E., X. QU, A. C. Jimenez-Vergara et al. 2011. Hydrogel-electrospun mesh composites for coronary artery bypass grafts. *Tissue Engineering* 17:451–461.

Miron, R. J. and Y. F. Zhang. 2012. Osteoinduction: A review of old concepts with new standards. *Journal of Dental Research* 91:736–744.

O'Brien, F. J. 2011. Biomaterials and scaffolds for tissue engineering. *Materials Today* 14:88–95.

Olson, J. L., A. Atala, and J. J. Yoo. 2011. Tissue engineering: Current strategies and future direction. *Chonnam Medical Journal* 47:1–13.

Phillips, B. L. and C. Callaghan. 2017. *The Immunology of Organ Transplantation. Surgery* (Oxford) 35:333–340.

Post, M. and C. Weele. 2014. Principles of tissue engineering for food. In *Principles of Tissue Engineering* (Eds.) R. Lanza, R. Langer, and J. Vacanti, pp. 1647–662. Tokyo, Japan: Academic Press.

Raffa, V., O. Vittorio, V. Pensabene, A. Menciassi, and P. Dario. FIB-nanostructured surfaces and investigation of Bio/nonbio interactions at the nanoscale. *IEEE Transactions on Nanobioscience* 7:1–10.

Ravichandran, R., J. R. Venugopal, S. Sundarrajan, S. Mukherjee, R. Sridhar, and S. Ramakrishna. 2012. Minimally invasive injectable short nanofibers of poly(glycerol sebacate) for cardiac tissue engineering. *Nanotechnology* 23:385102.

Rustad, K. C., M. Sorkin, B. Levi, M. T. Longaker, and G. C. Gurtner. 2010. Strategies for organ level tissue engineering. *Organogenesis* 6:151–157.

Sato, S. 2014. Mammalian cell culture system. In *Fermentation and Biochemical Engineering Handbook* (Eds.) C. M. Todaro and H. C. Vogel, pp. 17–24. Tokyo, Japan: William Andrew.

Sensharma, P., G. Madhumathi, R. D. Jayant, and A. K. Jaiswal. 2017. Biomaterials and cells for neural tissue engineering: Current choices. *Materials Science and Engineering C* 77:1302–1315.

Vacanti, J. and C. A. Vacanti. 2011. The history and scope of tissue engineering. In *Principles of Tissue Engineering* (Eds.) R. Lanza, R. Langer, and J. Vacanti, pp. 1–33. Tokyo, Japan: Academic Press.

Ventola, C. L. 2014. Medical applications for 3D printing: Current and projected uses. *Pharmacy and Therapeutics* 39:704–711.

Verma, A. 2014. Animal tissue culture: Principles and applications. In *Animal Biotechnology* (Eds.) A. S. Verma and A. Singh, pp. 211–231. Tokyo, Japan: Academic Press.

Verma, P. and V. Verma. 2014. Concepts of tissue engineering. In *Animal Biotechnology* (Eds.) A. S. Verma and A. Singh, pp. 233–245. Tokyo, Japan: Academic Press.

Zhang, X. and. Y. Zhang. 2015. Tissue engineering applications of three-dimensional bioprinting. *Cell Biochemistry and Biophysics* 72:777–782.

Zhu, W., X. Ma, M. Gou, D. Mei, K. Zhang, and S. Chen. 2016. 3D printing of functional biomaterials for tissue engineering. *Current Opinion in Biotechnology* 40:103–112.

7

Immobilization of Biomolecules

7.1 Immobilization

According to International Union of Pure and Applied Chemistry (IUPAC), immobilization is defined as, "the technique used for the physical or chemical fixation of cells, organelles, enzymes, or other proteins (e.g., monoclonal antibodies) onto a solid support, into a solid matrix or retained by a membrane, in order to increase their stability and make possible their repeated or continued use" (McNaught and Wilkinson 1997). In general, immobilization refers to restriction or retardation of molecule mobility and immobilized molecules are molecules that are attached to a solid structure and whose movement is restricted either partially or completely (Zhang et al. 2004). It is a natural phenomenon, for instance, microbial biofilms that are hydrated matrices with surface-attached microorganisms. Due to biological and biomedical application associated with biomolecules, they are common molecules to be immobilized (Kierek-Pearson and Karatan 2005). Some of the examples of immobilized biomolecules are shown in Figure 7.1.

7.2 Application of Immobilization

Different types of biomolecules are immobilized for different intended functions and applications. Immobilization of enzyme is widely practiced in biotechnological processes for cost reduction and efficient utilization of the enzymes. *Immobilized enzymes* refer to the physical confinement of enzymes in a certain defined space in such a way that the catalytic activities of enzymes are retained and facilitate repeated and continuous use (Tosa et al. 1966). Immobilized enzyme systems are used for production of regioselective and enantioselective compounds for biomedical applications (Lee et al. 2009a). Immobilized lipases are used for biosynthesis of polyesters

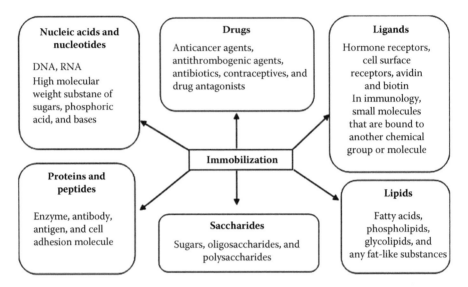

FIGURE 7.1
List of immobilized molecules.

(Idris and Bukhari 2011), and silica nanoparticles with immobilized laccase (oxidase enzyme) are used in wastewater treatment (Zimmermann et al. 2011). Other applications of immobilized enzyme include biorecognition element in biosensors, for instance, glucose oxidase (GO) immobilized on electrospun polyvinyl alcohol (PVA) and surface-modified carbon nanotubes used in glucose biosensors (Wen et al. 2011), enzyme horseradish peroxidase (HRP) immobilized on γ-aluminum trioxide nanoparticles/chitosan film-modified electrode used in hydrogen peroxide biosensors (Liu et al. 2011), and soluble plant invertase enzyme immobilized on composite of agarose-guar gum biopolymer matrix utilized in phenol biosensors (Bagal and Karve 2006).

The concern about biological toxin contamination in drinking water developed suitable and sustainable alternative to silver and titanium dioxide nanoparticles for water purification. Bionanomaterials such as apatmer, which is a single-stranded DNA or RNA molecule with specific sequences for molecular recognition selected from synthetic nucleic acid libraries, can detect the specific target molecules including protein, nucleic acids, amino acids, cells, and organic molecules. Apatmer covalently immobilized in nanomaterial, graphene oxide, was used to specifically recognize and adsorb peptide toxin produced by cyanobacteria in drinking water (Hu et al. 2012).

Immobilized biomolecules have revolutionized the field of drug delivery system with the benefits of target delivery and controlled release mechanisms. In drug delivery system, drug is released in response to stimuli such as pH or temperature and this mechanism is utilized in the development of glucose-sensitive insulin hydrogel beads. Insulin loaded in hydrogel beads swells at normal body pH (7.4) preventing release of insulin but shrinks at low pH (pH 4) as the blood glucose level increases and thus insulin is released in response to glucose concentration in the blood (Sona 2010, Elnashar 2010). Magnetic nanoparticles loaded with drugs are the novel techniques of drug delivery with the advantages of smaller size, larger surface area, magnetic response, biocompatibility, and nontoxicity. Other medical applications for diagnostics and therapeutic purpose utilizing immobilization techniques include development of enzyme-linked immunosorbent assays (ELISA), production of antibiotics such as penicillin G using immobilized cells (Carpentier and Cerf 1993), development of alternative for lactose intolerance utilizing immobilized β-galactosidase on thermostable biopolymers (Elnashar et al. 2009), and treatment of rheumatoid arthritis and joint disease. Nonmedical application of immobilized system includes treatment of wastewater contaminated with pesticide (Horne et al. 2002). Algal cells immobilized in either polymeric or biopolymeric matrix have great potential for application in wastewater treatment either by removing metal ion or nutrients from contaminated water. They are also utilized for the production of biohydrogen, biodiesel, and pigment. Other applications of immobilized algae involve fabrication of biosensors such as immobilized *Scenedesmus subspicatus* algal cell in optical biosensor for the determination of the herbicide content; immobilized *Chlorella vulgaris* in a biosensor for the detection of perchloroethylene aerosols; or toxic compounds such as atrazine, toluene, benzene, and immobilized *Scenedesmus capricornutum* used for testing toxicity of cadmium ions, copper ions, sodium dodecyl sulfate, pesticides, herbicides, and fungicides (Erogulu et al. 2015).

7.3 Supports or Matrix Used for Immobilization

The support or matrix holds the molecules temporarily or permanently and assists in immobilization of the respective molecules. Therefore, wise selection of support is paramount in determining the effectiveness of the immobilization process. An ideal support system should exhibit affinity toward molecules to be immobilized, chemical inertness, microbial resistance, mechanical stability, rigidity, feasibility of regeneration, and biodegradability (Foresti and Ferreira 2007). Mesoporous structures with large surface area and higher number of pores are one of the desired properties of the matrix. Such structure facilitates the high molecule loading per unit mass along with the better protection of the immobilized molecules from the surroundings (Lee et al. 2009a).

There are different types of matrices available, which may differ in their physical and chemical properties such as pore size and affinity toward water and chemical properties. Based on their chemical composition, matrices are divided into two major groups: organic and inorganic groups. Organic matrix is further classified into natural and synthetic polymers. Natural polymers such as alginate, chitosan and chitin, collage, carrageenan, gelatin, cellulose, starch, pectin, and sepharose are used as support for immobilization purpose. Synthetic polymers include insoluble ion exchange resins or polymers with porous surface, for example, diethylaminoethyl cellulose (DEAE cellulose), polyvinyl chloride (PVC), and UV-activated polyethylene glycol (PEG). Inorganic materials comprise of zeolites, ceramics, celite, silica, glass, activated carbon, and charcoal (Datta et al. 2013).

Recently, nanostructures such as nanoparticles, nanocomposites, nanofibers, and nanotubes are being focused for immobilization and stabilization. These nanostructures are being preferred to conventional support or matrix because of their inherent ideal characteristic, that is, large surface area and high mechanical properties, which permits effective immobilization and minimum diffusion problem. Nanotubes also permit easy separation and reusability by simple filtration technique or magnetic separation. Compared to the conventional process nanoparticle-based immobilization provided three important attributes: (1) facilitate synthesis of nanoenzyme particles in high solid content without the use of surfactants and toxic reagents, (2) particle size or nanoparticles can be tailored within effective working limits, and (3) attainment of homogeneous and well-defined core-shell nanoparticles with a thick enzyme shell (Ansari and Hussain 2012).

7.4 Methods of Immobilization

The immobilization techniques should be selected based on the molecules intended to be immobilized and the desired reaction to be achieved. The factors that need to be considered during appropriate method selection include process specification, reaction rate, enzyme deactivation and regeneration, cost of the immobilization, toxicity of reagents, and desired final characteristics of the immobilized compounds. Immobilization can be divided into two types: (1) physical and (2) chemical methods. Physical methods imply reversible entrapment of the molecules to the support based on physical forces: van der Waals force, hydrophobic interaction, and hydrogen bonding. The common physical method of encapsulation includes entrapment, adsorption, and microencapsulation. Chemical method of encapsulation involves irreversible attachment of the molecules to the support using covalent or ionic bonds. Covalent bond, cross-linking, ionic binding, and conjugation by affinity ligands are the chemical immobilization methods (Dwevedi 2016). The different types of immobilization techniques are illustrated in Figure 7.2.

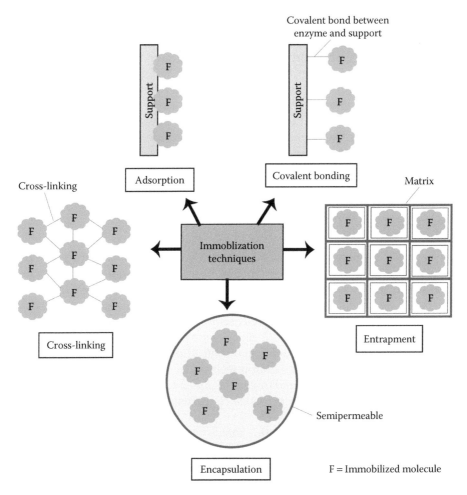

FIGURE 7.2
Different methods of immobilization. (From Zhao, X. et al., *Renew. Sustain. Energ. Rev.*, 44, 182–197, 2015. With Permission.)

7.4.1 Adsorption

In adsorption immobilization technique, the molecules are attached reversibly to matrix by relatively weaker noncovalent linkages including ionic or hydrophobic interactions, van der Waals forces, and hydrogen bonds without preactivation of support. The different types of support used may be inorganic mineral support (e.g., aluminum oxide, clay, activated carbon, bentonite, and porous glass) or organic support (e.g., starch, chitosan-modified sepharose, and ion-exchange resins). This immobilization technique requires optimization of variables such as pH, temperature, solvent nature, ionic strength, substrate, and support concentration (Alloue et al. 2008). Chemical modification of the conventional support improves the immobilization. For enzyme adsorption, silanized molecular sieves have been found to be better support because silanols on pore walls facilitate enzyme immobilization by hydrogen bonding. *Yarrowia lipolytica* lipase adsorbed on octyl-agarose and octadecyl-sepabeads were reported to have higher yield and greater stability compared to free lipase (Cunhan et al. 2008).

Adsorption immobilization method is a relatively simple technique with the advantages of low immobilization cost, retention of high enzyme activity, and chemical free enzyme binding. However, the strength of interaction between support and molecules is susceptible to surrounding conditions (Mohamad et al. 2015).

7.4.2 Entrapment

Entrapment immobilization method involves cross-linking of the molecule and polymer matrix in every direction either by covalent or noncovalent bond. The commonly used polymer matrix is polyacrylamide gel, cellulose triacetate, agar, gelatin, carrageenan, and alginate. This method permits continuous passage of substrate and release of the product from the matrix leading to the advantages such as continuity of the process, simplicity, sustainable enzyme properties, and no chemical modification. However, the major drawbacks of entrapment immobilization techniques are the chances of enzyme leakage, suitable only for small molecules, need to balance mechanical properties of the matrix and the presence of diffusional constraints. (Nakarani and Kayastha 2007).

Alginate-gelatin-calcium hybrid matrix used for enzyme entrapment has reduced the risk of enzyme leakage and improved mechanical stability. Nanostructured supports such as electrospun nanofibers and pristine materials have improved immobilization techniques. Entrapment of *Candida rugose* lipase enzyme in chitosan has reduced friability and leaching problem. Similarly, activity of lipase has been reported to be improved with entrapment by mesoporous silica and lipase entrapped in κ-carrageenan has been reported to have enhanced thermostability and organic solvent-resistant property (Datta et al. 2013).

7.4.3 Microencapsulation

In microencapsulation immobilization technique, membrane capsules made up of semipermeable polymers with controlled porosity are used to enclose the enzyme or bioactive compounds. The semipermeable membranes can be either permanent that is made up of cellulose nitrate and polystyrene or nonpermanent that is made up of liquid surfactant. Microencapsulated enzymes have relatively large surface areas that improve their catalytic efficiency (Dwevedi 2016).

7.4.4 Covalent Bonding

This is irreversible immobilization method that involves the formation of covalent bonds between the chemical group of both bioactive molecule and support. As hydroxyl groups and amino group of support or enzyme can form covalent bond, it is one of the widely used methods of enzyme immobilization. The chemical groups in the support or carrier that can form covalent bonds with support are amino, hydroxyl, carboxyl, thiol, guanidyl, imidazole, and phenol. For the formation and activity of covalent bond, the size, shape, and composition of the matrix and direction of enzyme binding are the important factors. Covalent bonds involve strong linkage between lipase and matrix, thus permitting reuse of enzyme and preventing enzyme leaching. This method also improves half-life and thermal stability of the enzymes when coupled with different supports such as mesoporous, silica, and chitosan (Agyei et al. 2015).

Cross-linking of enzymes with electrospun nanofibers has shown greater residual activity due to the increased surface area and porosity. Use of such nanodiametric supports for enzyme immobilization has revolutionized the immobilization sector. Covalent binding of alcohol dehydrogenase on attapulgite nanofibers (hydrated magnesium silicate) has been opted owing to its thermal endurance and variable nanosizes. Biocatalytic membranes have been useful in unraveling effective covalent interactions with silicon-coated enzymes. Covalently linked enzymes on magnetic nanocluster have potential application in medical sector due to their operation stability and reusability (Datta et al. 2013).

7.4.5 Cross-Linking (Copolymerization)

This involves the formation of number of covalent bonds between the enzyme and matrix using bi- or multifunctional reagents (e.g., glutardialdehyde, glutaraldehyde, glyoxal, diisocyanates, hexamethylene diisocyanate, toluene diisocyanate etc.). Glutaraldehyde (GA) is the most commonly utilized cross-linking agent as it is easily available and cheap. Generally, amino groups of lysine, sulfhydryl groups of cysteine, and phenolic OH groups of tyrosine or imidazol group of histidine are used for enzyme binding under mild conditions. The main advantage of this method is its simplicity. However, it leads to the loss of large amount of enzyme due to the nonregulation of the reaction. Further, this method of enzyme immobilization also suffers limitation caused by diffusion (Mulagalapalli et al. 2007).

7.5 Surface Immobilization

It is essential to select suitable surface for immobilization to result in accurate, precise, and sensitive bioassays. The surface acts as an interface with the biological components and performs important function to immobilize the molecule resulting in optimal biomolecular activity remaining inert to the reaction. Surface immobilization of biomolecules has wider application in the development of biosensors.

Enzymes glucoamylase, β-D-fructofuranosidase, and β-D-glucosidase were surface immobilized on GA-cross-linked gelatin beads. This method can be coupled with other immobilization technology for the production of gelatin-entrapped enzymes. Dual immobilized enzyme system with glucoamylase and invertase has been developed. Coupling of glucoamylase onto cross-linked gelatin particles by precipitation with polyhexamethylenebiguanide hydrochloride has also been designed (Kennedy et al. 1984).

Surface immobilization of the neural adhesion molecule L1 on neural probes showed improved functionality of the implanted neural electrodes. Immobilized DNA aptamer (synthetic functional oligonucleotide receptors) on the gold surface along with oligoethylene glycol thiol, when tested against immunoglobulin E and vascular endothelial growth factor as target protein, proved surface immobilization as a versatile strategy for biosensing and protein interaction analysis (Zhang and Yadavalli 2011).

Taunk et al. (2016) studied surface immobilization of quorum-sensing biofilm inhibiting furanones (FU) and dihydropyrrol-2-ones (DHPs) onto azide-functionalized glass surface by photoactivation, which showed antibacterial activity against *Staphylococcus aureus* and *Pseudomonas aeruginosa*. Hence it was observed that surface immobilization can be an effective technique to produce novel antibacterial biomaterial surfaces.

7.6 Gel Immobilization

Solid supports that are either organic or inorganic, such as silica, clay, and collagen, for enzyme immobilization are considerably studied due to their efficiency in enzyme stability. Among the several available supports, use of silica gel is extensively studied. Size of silica particles is similar to the nano-biomolecules, that is, in the range of nanometer scale and exhibits the properties such as high surface area, regular structures, high stability to chemical and mechanical forces, and resistance to enzyme attack. Immobilization of protease on modified silica gel (aminated by 3-aminopropyltriethoxysilane) by GA cross-linking showed retention of enzyme activity for more than 40 days (Nazari et al. 2016).

Furthermore, sol–gel immobilization process can be also applied to immobilized living cells by forming a porous gel network around cells. Mycelium and pellet of *Aspergillus oryzae* were bioencapsulated in sol–gel network formed by hydrolysis of inorganic compound, tetraethyl orthosilicate. This hydrolysis generated inert glasses with high porosity, and high thermal and mechanical resistance. The sol–gel matrix can be reinforced by hybridizing with other molecules such as starch, gelatin, methyl methacrylate, chitosan, polyacrylamide, calcium alginate, or polyethylene oxide (organic compound). According to many reports, nonswelling inorganic silica materials with soft calcium alginate provide several advantages associated with strength and thermostability. α-amylase activity was found to increase in *A. oryzae* immobilized by 10% (w/v) calcium alginate by sol–gel immobilization technique. According to the result, immobilization of microbial cells in silica-based matrices has been proved to be a good strategy to enhance the biosynthetic capabilities (Evstatieva et al. 2014).

7.7 Immobilization of Bioactive Compounds

Bioactive compounds are defined as essential and nonessential compounds occurring naturally in foods and are shown to have an action on human health. They are also known as nutraceuticals and described as natural constituents in food with associated health benefits beyond the basic nutritional value (Biesalski et al. 2009). Some of the common bioactive compounds that have potential health benefits include proteins and peptides; phenolic compounds; and flavonoids such as quercetin, catechin, isoflavones, lycopene, soluble dietary fiber, plant sterol, and carotenoids. These compounds may be isolated from plants, animals, microalgae, microbial sources, or synthetically produced sources (Kris-Etherton et al. 2002).

Peptides and proteins as bioactive compounds exhibit wide range of health benefits including antihypertensive and immunomodulatory activities. However, bitter taste, loss of *in vivo* bioactivity, and poor stability associated with proteins and peptides need to be overcome to expand their utility in human nutrition. Different encapsulation and immobilization techniques such as entrapment in matrix of liposome, cyclodextrin, and hydrogel are the suitable ones to improve the stability, solubility, and bioavailability of bioactive compounds. Riboflavin was immobilized in whey microbead matrix formed by cold-set gelation technique that is suitable for heat-sensitive bioactive compounds. In cold-set gelation technique, gelation is induced by ions, for example, heat-denatured whey proteins are cross linked with a divalent ion (calcium ion) (O'Neill et al. 2016).

7.8 Immobilization of Live Cells

Self-immobilization of cell is the natural phenomenon like the slimy layer resulting from the growth of cells on surfaces as seen in microbial biofilms. However, the cells are being immobilized intentionally with the aim of increasing volumetric productivity and product concentration, and decreasing the substrate concentration in the product. Usually enzymes are immobilized as they show higher enzyme activity and stability but immobilization of whole cells is used as an alternative in the cases in which the enzyme loses its activity when immobilized or when the enzyme purification cost is high. Whole cell immobilization refers entrapment of the intact cells to a certain defined region with the preservation of desired catalytic activity (Karel et al. 1985).

Cell-based therapeutic strategy has been around for more than two decades now. Primary cells or cell lines with specific hormonal or bioactive compound secretions can be used as alternative therapeutics in various chronic diseases and/or deficiencies such as *Diabetes mellitus type 1* and in neurological disorders such as *Parkinson's disease, Alzheimer's disease* (Skinner et al. 2009; Teramura and Iwata 2010). Transplantation of primary or genetically engineered cells of allo- or xenogenic organ has emerged as a promising approach for the localized and regulated *de novo* delivery of a nearly unlimited number of therapeutic agents with potential capacity to cure or treat virtually any disease or disorder, not limiting toward the endocrine system (diabetes, hyperparathyroidism, adrenal insufficiency), central nervous system (Parkinson's, Alzhemeier's, Huntington's) as well as cancer, kidney failure, and cardiovascular diseases (Zhou et al. 2005; Skinner et al. 2006; Wollert and Drexler 2006; LÖhr et al. 2006). However, widespread clinical application of cell transplantation remains limited, in part, by the deleterious side effects of current immunosuppressive regimens necessary to prevent host rejection of transplanted cells. It has long been postulated that systemic immunosuppression could be eliminated if transplanted cells and tissues were physically isolated from the host immune system using a semipermeable membrane or immunoisolation device (Muschler et al. 2004). The novel idea for developing the artificial pancreas was generated by Chick et al. (Chick et al. 1975) and later Lim and Sun developed the alginate-encapsulated islet cells technology (Lim and Sun 1980). The extensive research has focused on the design, fabrication, and application of immuno-isolation devices capable of protecting transplanted allo- and xenogenic cells from the host while facilitating adequate transport of oxygen, nutrients, and secreted therapeutic molecules.

Protection of transplanted cells from deleterious host immune responses is necessary to exploit the full clinical potential of cell-based therapeutics, and immunoisolation stands to play a pivotal role toward this end. Maintenance of adequate oxygen and nutrient transport is of paramount importance for success of any transplant, but particular attention must be given to immuno-isolated cells due to limited vascularization of the device or at the intended site of transplantation. Early inflammatory responses are intrinsic to any implanted cell material composite, and such responses, if not inherently detrimental, can initiate chronic inflammation or otherwise intensify immune responses to transplanted cells and tissues. Recent advances in immunobiology have identified fissures in the prevailing of dogma of immunoisolation, particularly as applied to xenogenic tissue. Xenogenic cells are those transplanted across species barriers, whereas allograft cells are those transplanted within the same species.

Immunoprotection of xenogenic cells requires several stringent biological and physical criteria to be met. In live cell therapy, xenogenic or allogenic cells are needed to confine within a carefully fabricated biocompatible immunoisolating device (Chang 1964; Opara et al. 2002). The device containing the living cells are then transplanted into the host. The objective of transplanting the foreign cells or tissue is to replace lost function in the host tissue due to disease or degeneration. The principle of immunoisolation of cells for transplantation has two major potential benefits: (1) cell transplantation without the need for immunosuppression and its accompanying side effects and (2) transplantation of cells from xenograft to overcome the limited supply of donor cells (Beck et al. 2002; Portero et al. 2010). The immunological responses to a xenograft have various features that distinguish from the alloresponse. The first difference depends on the species crossed in which the recipient may have natural cytotoxic antibodies against the xenoantigens in the donor tissues (Lanza et al. 1991; Cole et al. 1992; Orive et al. 2006). In addition, the major method of recognition of xenoantigens is through the indirect pathway in which xenoantigen is processed and presented by the host antigen-presenting cells (APCs) to the Th- cells.

Protection of transplanted cells from the host immune system using immunoisolation technology is important in realizing the full potential of cell-based therapeutics. Microencapsulation of cells and cell aggregates has been the most widely explored immunoisolation strategy, but widespread clinical application of this technology has been limited, in part, by inadequate transport of nutrients, deleterious inflammatory responses, and immune recognition of encapsulated cells through antigen-presenting pathways (Wang et al. 1997; Tam et al. 2011). For immunoprotection of these live cells, the polymeric membrane should allow permeability of glucose, insulin, oxygen, and other metabolically active products to ensure the functionality and therapeutic effectiveness of those cells (Kizilel et al. 2005; Lee et al. 2009b). However, it must also prevent the passage of cytotoxic cells, macrophages, antibodies, and complement to remain effective. The successful immunoprotection requires the membranes that not only provide protection of the encapsulated cells from the host immune system but also have properties that diminish the release of xenogenic antigens (those released from foreign tissue and are typically <20 nm) through reduced pore size and surface properties that prevent nonspecific adsorption (Sakai et al. 2001; Krol et al. 2006; Wilson et al. 2008). Biocapsules with membrane pores in the 10 s of nm range seem suitable for application in xenotransplantation (Desai et al. 2000). The molecules having less than 35 Å, such as insulin, glucose, oxygen, and carbon dioxide pass easily through these membranes. The immunoglobulins and other antigens are generally bigger than 10 nm. Many polymeric devices have been developed for the purpose of immunoisolation of transplanted cells secreting hormones, neurotransmitters, growth factors, and other bioactive cellular secretary products (Visted et al. 2001; Narang and Mahato 2006).

7.8.1 Immobilization of Probiotics

Based on understanding of the mechanisms of action of probiotics in human health, probiotics have different definitions. Commonly, probiotics are known as live microbial feed supplements that have beneficial effect on host leading to improved intestinal microbial balance. According to Food Agriculture Organization (FAO) and World Health Organization (WHO), probiotics are defined as "Live microorganisms (bacteria or yeasts), which when ingested or locally applied in sufficient numbers confer one or more specified demonstrated health benefits for the host" (Anal and Singh 2007). Probiotics colonized in the large intestine supplements the natural flora of the gastrointestinal (GI) tract with additional bacteria, which perform different functions such as alleviation of intestinal disorders, stimulation of immune system, inhibition of carcinogenic compounds, reduction of toxic metabolites, serum cholesterol, maintenance of mucosa integrity, alleviation of lactose intolerance, and prevention of vaginitis (Fang et al. 2012).

Lactic acid-producing bacteria, also known as lactic acid bacteria (LAB), are the most important probiotic microorganism associated with the human GI tract. LAB produce lactic acid via fermentation through cell-recycle repeated batch. Under batch fermentation process, lactic acid production by free sensitive cells is inhibited by substrate and end product accumulation leading to lower production efficiency and increased downstream cost. Lactic acid-producing bacteria, *Lactobacillus rhamnosus*, was immobilized in mesoporous silica-based matrix with high immobilization efficiency of 78.77% and high glucose conversion yield of 92.4%. The immobilized cells showed stability during repeated fermentation processes (8 batches) with no reduction in lactic acid production (Zhao et al. 2016).

Yogurt is produced from milk by the action of LAB such as *Lactobacillus delbrueckii ssp. bulgaricus* and *Streptococcus thermophilus*. Fortification of yogurt with combination of prebiotic and probiotic bacteria such as *Bifidobacteria* and *Lactobacilli* will improve the nutritional value. However, incorporation of probiotic in yogurt is a challenge due to their interaction with natural yogurt microflora and yogurt compositions, processing, and storage conditions (pH, temperature, lactic acid concentration, oxygen, micronutrients, etc.), which may lead to loss in cell viability. To obtain the benefit from consumption of probiotic yogurt, it is important to maintain the minimum level of viable probiotic cells in the range of 10^6–10^9 cfu/ml to survive the extreme acidic condition under the GI tract. In such cases, immobilization of cell is a new method that has potential to maintain probiotic cell viability during storage period and delivery through the GI tract. Probiotics (*Lactobacillus casei* ATCC393 and *Lactobacillus bulgaricus* DSM20081) were immobilized by wheat bran (*Triticum aestivum*) as a cell immobilization carrier, and novel yogurt was produced that showed significantly higher viable cell compared to traditional product with free cell under storage at 4°C (Terpou et al. 2017).

7.8.2 Encapsulation Technology in Cell Transplantation

Encapsulation is defined as a technology of packaging solids, liquids, or gaseous materials in miniature, sealed capsules that can release their contents at controlled rates under the influences of specific conditions (Anal et al. 2003; Anal and Stevens 2005; Anal 2008). Many encapsulation procedures have been proposed but adopting those procedures universally for drugs, cells, and other bioactive components in pharmaceutical and biomedical devices is still a challenge. Most of these drugs, cells, and bioactive compounds have their own molecular characteristics (e.g., molecular weight, polarity, solubility etc.) and exposure to the immune systems, which render them to imply different encapsulation approaches.

Encapsulation of cells provides the means of transplanting cells in the absence of immunosuppressive drugs. In diabetic patients, transplantation of pancreatic islet cells could restore normalglycemia (Elliott et al. 2007). However, as with most tissues or cellular transplants, the islet grafts, particularly xenografts, are subjected to immunorejection in the absence of chronic immunosuppression. To overcome this need for immunosuppressive drugs, the concept of isolating the islets from the recipient's immune system within biocompatible size-based semipermeable capsules was developed (Desai et al. 2000). Bioencapsulation is a physical process where thin films or polymer coats are applied to small solids, liquid droplets, or gaseous materials in miniature. These sealed capsules that can protect from external environment release their entrapped contents at controlled rates under the influences of specific conditions (Anal and Singh 2007). Bioencapsulation has been shown to be efficacious in mimicking the cells' natural environment and thus improves the efficiency of production of different metabolites and therapeutic agents (Tziampazis and Sambanis 1995; de Vos and Marchetti 2002). A microcapsule consists of a semipermeable, spherical, thin, and strong membrane surrounding a solid/liquid/gas core, with a diameter varying from one micrometer to one millimeter. The principle of encapsulation is that the transplanted cells are contained within an artificial compartment separated from the immune system. Thus, the capsules should protect the cells from potential damage caused by antibodies, complement, and immune cells (Dionne et al. 1996; Li et al. 1996). However small molecules such as nutrients should be able to freely enter through the capsular membrane, whereas the waste products and hormones such as insulin should be easily released.

Despite significant progress in cell transplant therapy over the last four decades, long-term and complete immunoisolation of xenogenic cell grafts via membrane encapsulation remains a much sought-after therapeutic goal.

The encapsulating and coating materials chosen for a particular application would depend on the physical, chemical, and biological properties of the core materials as well as the method used for capsule formation. These materials must be nonreactive with the core materials, biocompatible, and biodegradable. Coating materials, which are basically film-forming materials, can be selected from a wide variety of natural or synthetic polymers, depending on the material to be coated and characteristics desired in the final encapsulated products. The composition of the coating material is the main determinant of the functional properties of the microcapsules and of how it may be used to improve the performance of a particular ingredient (Anal and Singh 2007). An ideal coating material in biomedical and pharmaceutical applications should exhibit the following characteristics: (1) good rheological properties at high concentrations and easy-to-work-with property during encapsulation, (2) ability to disperse or emulsify the core materials and stabilize the emulsion or particles produced, (3) nonreactivity either physically or chemically with the materials to be encapsulated both during processing and on prolonged storage, (4) ability to seal and hold the core materials within its structure during processing and storage, (5) ability to provide maximum protection to the active material against environmental conditions, (6) immunoprotective, and (7) inexpensive and biocompatible (Li 1998; de Vos et al. 2003; Anal 2008).

Various properties of micro/nanocapsules that may be changed to suit specific ingredient applications include composition, mechanism of release, particle size, final physical form, and cost. Before considering the properties desired in encapsulated products, the purpose of encapsulation must be clear. In designing the encapsulation process, the following questions should be taken into consideration (Bhopatkar et al. 2005; Nafea et al. 2011):

- Functionality of the encapsulated ingredients and the final products
- Biocompatibility of the coating materials
- Processing conditions that do not destroy the properties of cells
- Optimal concentrations of cells (core substances) and dosage forms
- Mechanism of release of core substances and bioactive components
- Particle size, density, and stability requirements
- Cost constraints of the encapsulated ingredients and encapsulating materials

7.8.3 Immunocompatibility and Biocompatibility of Materials Used for Encapsulation

The immunoprotection of transplanted cells offers transplantation without the need of immunosuppression of the host. Moreover, immunoisolation of a xenograft, for example, a transplant of cells from nonhuman species, has been suggested as a way of overcoming the limited supply of human organs for transplantation (O'Shea and Sun 1986; Lum et al. 1991; Lum et al. 1992; Wang et al. 1997). The membranes used for encapsulation of living cells must be biocompatible and immunocompatible to the recipient and to the cells it encloses. The materials used in biomedical and pharmaceutical products reflect complex characteristics that influence the fate of the material in its biological host. However, the ability of the host to accept the material depends not only on the material itself but also on the host, its genetic depositions, and physical status (Granicka et al. 1996; Lanza et al 1999). The device used for encapsulation should not evoke fibrous tissue reaction, macrophage activation, and cytotoxic agent release (Rihová 2000; Lee et al. 2009a).

7.8.4 Bioencapsulation by Polyelectrolyte Complexation

Interaction of oppositely charged polymers is one of the simplest forms of physical membrane barrier around living cells. Being highly hydrophilic charged polymers offers the feasibility of developing an aqueous encapsulation system that is compatible with cellular environment. The complexation between polyanionic polymers with polycationic polymers has been widely used for encapsulating a variety of bioactive compounds and cells. A major pursuit has been the encapsulation of pancreatic islets for the treatment of Type-1 diabetes.

The bioencapsulation technique is based on the entrapment of individual islets in an alginate droplet, which is transformed into a rigid bead by gelification in a divalent cation solution, such as in calcium chloride solution (Anal and Stevens 2005). Alginic acid and their derivatives have also been successfully used by the pharmaceutical industry for over 40 years. Their unique properties make them effective in a variety of applications including tablet disintegration, controlled release, encapsulation, films and coatings, lubricating agents, prevention of gastric reflux, gelling, and thickening agents to stabilize emulsions and suspensions. Alginates are the most frequently employed biomaterials for cell immobilization due to their abundance, easy gelling properties, and apparent biocompatibility.

Alginic acid, a natural polymer, is a polyuronic acid extracted from seaweeds and is composed of varying proportions of 1-4 linked β-D-mannuronic (M) and α-L-guluronic acids (G). These residues are present in varying proportions depending on the sources of the alginic acid. It turned out that alginic acid (and its salts) is a block copolymer, containing both MM…and GG…homopolymer blocks and mixed blocks containing irregular sequences of M and G units (Atkins et al. 1973; Annison et al. 1983). The binding of divalent cations and subsequent gel formation is dependent on the composition and arrangement of the blocks of residues. In particular, gel strength is related to G content. The proportion of these segments varies between each species of kelp and imparts distinctly different properties to the final product. Depending on the specific species of kelp used in manufacturing, ratios of mannuronic acid to guluronic acid contents (M/G ratio) typically range from 0.4 to 1.9 (Orive et al. 2005; Orive et al. 2006). The GG blocks have preferential binding sites for divalent cations, such as Ca^{2+}, and the bound ions can interact with other GG blocks to form linkages that lead to gel formation. On addition of sodium alginate solution to a calcium solution, interfacial polymerization is instantaneous with precipitation of calcium alginate followed by a more gradual gelation of the interior as calcium ions permeate through the alginate (Qi et al. 2008). Other than calcium, barium has also been tried and explored as cationic divalent for gelification of alginate. As barium is known to be toxic, concerns have been raised to using this ion as a cross-linking agent (Wikström et al. 2008).

Alginate provides some major advantages over the other systems. First it has been found, repeatedly, not to interfere with cellular function of live cells (de Vos et al. 1996; Tam et al. 2006). Alginate is one of the few materials that allow for processing of the capsules at physiological conditions. The encapsulation can be done at room temperature, at physiological pH, and in isotonic solutions. Also, it has been shown that alginate capsules can provide a microenvironment that facilitates functional survival of islets and other live cells (Wilson and Chaikof 2008). It has been postulated that the three-dimensional matrix provides a growth support for the islets and also prevents clumping and fusion of the free islets, which can interfere with the availability of nutrients and oxygen for the islet cells in the core of the clumps. Also, alginate-based capsules have been shown to be stable for years in small and large animals and also in human (Soon-Shiong et al. 1992).

The attachment of the cells to the surface of the alginate microcapsules is limited due to its negative charge. Crude alginate processed from seaweed contains polyphenols, proteins, and endotoxins (Kluseng et al. 1999). Hence, the purification of these types of polymers is must before applying in any such immunocomprised applications. Formation of alginate-based immuno-isolating capsules requires alginate of extreme purity, particular range of viscosity, and standardized mannuronic/glucuronic acid composition. As a natural polymer, alginate is limited by its tendency to be largely contaminated. Purification of crude commercial alginate is necessary before using this material for encapsulating the cells. Previous techniques employed to reduce contaminants (e.g., proteins, endotoxin) rely on organic solvent gradients to selectively solubilize and remove them throughout a pH range to maintain alginate as a precipitate. Such techniques are highly effective at removing protein but are often associated with secondary effects including molecular weight portioning, conformational changes, and residual solvents, all of which ultimately alter the chemical behavior of alginate in the aqueous solutions. Endotoxin can similarly be removed by manipulating its tertiary structure to form large aggregates that exceed the pore size of the filters used for purification process. In addition, materials such as diatomaceous earth are used to selectively bind the lipid moiety of endotoxin in combination with manipulation of shape by creating a pH gradient in the filter environment. This unique technique would remove almost all polyphenolic compounds, proteins, and lipopolysaccharides associated with alginates. The safety of alginate in biomedical and pharmaceutical applications in tissue-engineered medical products (TEMPS) should be established and certified in accordance with current guidelines such as ISO 10993 and practice ASTM F748. Even though a number of purification procedures, especially for the removal of polyphenolic compounds and the antigenic proteins, have been described (Skjak-Break et al. 1989; Klöck et al. 1994), this still remains a major challenge in developing a truly purified commercial ingredient. Polyphenols are known to be toxic to cells and responsible for the oxidative–reductive depolymerization (ORD)-catalyzed depolymerization of alginates. Endotoxins are potent immunostimulators. Hence, the purification of alginate is must before applying in any biomedical applications. With the advanced technology, there is a possibility of purification of alginates with the very low level of endotoxin (less than 100 EU/g). The improvement in biocompatibility has been found with the purified alginate (de Vos et al. 1997, 2003).

Alginate-based microcapsules have been applied for immunoisolation as coated and uncoated beads (King et al. 1987; Siebers et al. 1992; Grohn et al. 1994; Klöck et al. 1997; Hasse et al. 1998; Darrabie et al. 2005; Wang et al. 2009). Calcium-alginate microcapsules have the disadvantage that they are sensitive to chelators such as citrate, phosphate, and lactate (Anal and Singh 2007). Thus, long-term survival of solid Ca^{2+} cross-linked alginate microcapsules is limited, but this can be advantageous when autologous cells are transplanted, such as chondrocytes and osteoblasts for the restoration of cartilage and bone

(King et al. 1987; Hasse et al. 1998). The chelators sometimes are used with Ca^{2+}-based microcapsules to produce liquid core (Hasse et al. 1998). In past, uncoated barium-alginate beads were tried to encapsulate the islet cells and transplant to the animals. A study by Duvivier-Kali et al. (2001) reported that normoglycemia could be achieved in all streptozotocin (STZ)-induced diabetic and spontaneously diabetic NOD mice transplanted with syngeneic and allogeneic islets enclosed within this barium-alginate membrane. The study and some other reports suggested the potential of these systems. The uncoated barium-alginate microcapsules had a molecular weight cut-off of 600 KD, whereas immunoglobulin G (IgG), the smallest of the immunoglobulins, has a molecular weight of 140 kD and most of the potentially harmful cytokines' molecular weight are in the range of 17.5 (IL-1) to 51 kD (TNF-α). The inward and outward permeability of the molecules through microcapsule membranes depends on the three-dimensional size, such as radii of gyration and the charge of both the molecules of interest and the polymer network in addition to the pore size and pore size distribution (Orive et al. 2006).

To overcome these problems, calcium-alginate microcapsules have been coated with a polycation in view of having broader potential application than calcium-alginate microcapsules because such coats increase the mechanical stability and a further restriction in permeability of immunoglobulins and cytokines. The most commonly used alginate-based microcapsules are coated with poly-L-lysine (PLL), but also other poly-L-ornithine (PLO), poly-D-lysine, chitosan, and polymethylene-co-guanidine have been investigated for their efficacy. After gelification of the capsules in calcium bath, the capsules are subsequently coated with the polycation membrane by suspending the microcapsules in polycation solutions such as PLL. During this step, polycations bind to alginate molecules (Koch et al. 2003; Bunger 2003; Zhang et al. 2008) and induce the formation of complexes at the capsule surface (Clayton et al. 1991; Thu et al. 1996), which causes the nanoporous membranes of microcapsules. Thickness of the microcapsules and the PLL barriers interferes with the delivery of immunogenic molecules such as cytokines and immunoglobulins.

7.8.5 Enhancing the Performance of Microencapsulated Islets and Implantation of Encapsulated Islet Cells in Animals and Clinical Trials

The need for combining more than one mechanism for enhancing the outcome of islet transplantation has been well recognized. Several researchers have evaluated the benefit of combining islet encapsulation with one or more additive strategies to improve the survival and functioning of the transplanted cells.

There have been few reports of the clinical application of encapsulated islets. The alginate-PLL-alginate capsules containing islet cells implanted in peritoneal cavity of rats were found in vast majority of intact capsules even after retrieving of two years (Thu et al. 1996). Of the retrieved capsules, only 2%–10% was overgrown with inflammatory cells, whereas 90%–98% of the alginate-PLL capsules were completely free of any inflammatory overgrowth. Tam et al. (2009) studied on the adsorption of human immunoglobulins (e.g., I_gG, I_gM, and I_gA) to implantable alginate-PLL microcapsules. The adsorption of these immunoglobulins was not affected by the types and purity of the alginates but highly dependent on the presence of the polylysine membrane, indicating that the PLL was mainly responsible for the adsorption. While other researchers have observed that the microcapsules prepared of purified alginate and intermediate-glucuronic acid, are highly biocompatible for prolonged periods of time when implanted in rats (de Vos et al. 1996, 2003). It is also known that microcapsules composed of alginates that are not properly purified induce a severe inflammatory response (King et al 1987).

Poly-L-Ornithine (PLO) has recently been applied successfully in animals and even for clinical applications. It is preferred over alginate-PLL-alginate capsules due to their more chemical stability but also immunoselective in terms of nominal membrane's molecular weight cut-off and also biocompatibility (de Vos et al. 2002; Thanos et al. 2006). To make a homogenous and biocompatible hydrogel, PLO needs to be ionically complexed with a mannuronic acid-enriched alginate. Capsules prepared from this alginate are very resistant to mechanical burst and the only way to dissolve them in is by exposuring it to strong bases. The intact capsules were retrieved unbroken and without much fibrotic growth when implanted and retrieved from the intraperitoneal cavities of rodents, canines, or pigs. Thanos et al. (2006) implanted the capsules made from five different purified and unpurified alginates in rats' intraperitoneal cavity over three months to evaluate the stability and fibrotic growth. Most of the capsules made from high-grade pure alginates remained intact without any fibrotic growth. Fourier transform infrared spectroscopy (FTIR) analysis and surface morphology from microscopy indicated that the enhanced alginate-polycation biocapsules have the capability of survival in all sites, including the harsh peritoneal environment, for at least 215 days. Elliot et al. (2005b) have tested some alginate-based encapsulated piglet islet formulations into diabetic mice and monkeys and noted the improvement of concentration of insulin in blood. In another study with a placebo-controlled design (Elliot et al. 2005; Calafiore et al. 2006), the safety and clinical trials of encapsulated porcine islets in alginate-based microcapsules were conducted in a nonhuman primate model of streptozotocin-induced diabetes. The disease was found worsening in case of control, whereas the insulin dose was needed to decrease in the animals transplanted with encapsulated islets. Though individual blood glucose values varied, one monkey transplanted with the encapsulated islets was weaned off insulin for 36 weeks. Another recent study reports that the use of intraperitoneally implanted encapsulated allografts in the type 1-diabetic patients remained nonimmunosuppressed but were unable to withdraw exogenous insulin (Elliott et al. 2007).

7.9 Immobilization in Biosensor Fabrication

Biosensor is an analytical device that functions to convert observed response into a measurable signal with the help of biological sensing component connected to a transducer. The signal generated is directly proportional to the concentration of biochemical or chemicals present in the sample (Pundir 2015). The major components of biosensor are bioreceptor, transducer, amplifier, and microelectronic as illustrated in Figure 7.3.

It is useful in fast, specific, and sensitive detection for the presence and concentration of biochemical compounds such as proteins, nucleic acids, drugs, cells, and ions. Basically it is composed of three parts: a bioreceptor that reacts with the specific target, transducer that converts target–bioreceptor reaction into a measurable signal, and a reader to interpret the signal output. This device has wide scope in the field of medical science, food, and environmental monitoring. In medical science, use of enzyme-linked immunosorbent assay (ELISA) is regarded as the gold standard in research and clinical diagnostics due to its high standard configuration, readability, and specificity. Nanomaterials such as nanotubes, gold, and silver, quantum dots, lipid bilayer modified with bioreceptors such as nucleic acid, peptide, biotin, polymer, antibody, and protein are being utilized for sensitivity assay (Mazur et al. 2017).

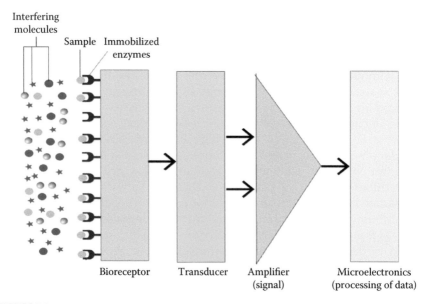

FIGURE 7.3
Schematic representation of working principle of biosensor. (From Pundir, C.S., *Enzyme Nanoparticles*. William Andrew, Tokyo, Japan, 2015. With Permission.)

When enzyme/protein molecules with diameter around 10 nm are aggregated together in the size range up to 100 nm, they exhibit specific optical, electrical, thermal, chemical, mechanical, and catalytic activity and this aggregate is known as enzyme nanoparticles. Different enzymes such as HRP, GO, cholesterol oxidase (ChOx), and uricase nanoparticles have been immobilized by desolvation and GA cross-linking techniques. These enzymes or protein when directly adsorbed on metal surface are prone to lose their bioactivity during reaction. But when they are immobilized onto metal nanoparticles and electrodeposited onto bulk electrode surface, their bioactivity is retained for longer period. Immobilized enzyme nanoparticle, immunoreagents, DNA, or whole cells are being used as bioreceptors in the development of improved biosensors. Immobilized enzyme nanoparticle-based biosensors can be either dissolved oxygen (DO) metric as in oxidoreductase-based biosensor that consumes oxygen (oxidase) and generates hydrogen peroxide or it can be amperometric in which immobilized enzyme generates reduced form of NAD(P)H, for example, dehydrogenase during the oxidation of target. Both DO and amperometric biosensor based on immobilized enzyme nanoparticles can be used for the detection of hydrogen peroxide, glucose, cholesterol, and uric acid in samples (Pundir 2015).

Different immobilization techniques that are suitable for specific enzyme for particular function are being studied. The adsorption, cross-linking, entrapment, and coating are the common method of enzyme immobilization. However, cross-linking is the popular method for fabricating glutamate sensors. In a research work by Tseng et al. (2014), bovine serum albumin (BSA) was used as reagent in glutamate biosensor and enzyme glutamate oxidase (GlutOx) was immobilized on platinum microelectrodes by two methods: (1) adsorption of GlutOx on electrodeposited chitosan and (2) cross-linking of GlutOx with GA as shown in Figure 7.4. Glutamate biosensor with glutamate oxidase immobilized by adsorption on chitosan showed faster response time

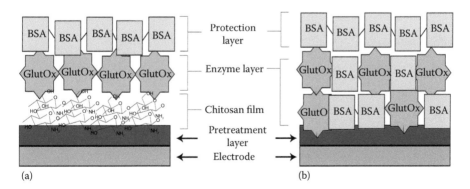

(a) (b)

FIGURE 7.4
Representation of the glutamate sensor with glutamate oxidase (GlutOx) immobilized by (a) adsorption on chitosan film and (b) cross-linking with stabilizing reagent BSA and cross-linker glutaraldehyde. (From Tseng, T.T.C. et al., *Molecules*, 19, 7341–7355, 2014. With Permission.)

and lower detection limit, whereas the biosensor with enzyme cross linked with GA showed larger linear detection range and higher sensitivity. *In vivo* study of the biosensor was also done by implanting the biosensor on rat's brain and monitoring the stress-induced glutamate release.

Nucleic acid biosensor constitutes single-stranded DNA/RNA as sensor probe, which can detect the specific target strands based on their ability to bind with complementary strands. The overall performance of sensor can be improved by immobilization of the nucleic acid sensor probes in such a way that sensor probes can detect target sequence by specific recognition signal, that is, sequence-specific hybridization interactions, whereas nonspecific interactions such as nucleobase–substrate interactions are minimized. Immobilization of sensor probe can be done by two ways: (1) covalent binding of one end of probe onto solid surface such as binding of 5′-thiol-modified DNA oligonucleotides in gold via gold–sulfur interaction or binding of 5′-amino-modified DNA oligonucleotide onto an epoxy-modified surface and (2) noncovalent binding such as binding of 5′ end of biotinylated nucleic acid is directly attached to carbon-based surface. Moreover, synthetic nucleic acid analogs such as peptide nucleic acid (PNA) are being used as sensor probes in electrochemical, optoelectronic sensor, and microarray-based sensor due to its peptidic and nonionic backbone. Use of peptide nucleic acid (PNA) sensor probe in nucleic acid sensor has advantages such as rapid PNA–DNA duplex formation and cheaper cost due to high sensitivity compared DNA–DNA duplex assay. The PNA probe method for detection of target nucleic acid strand by electrochemical contains four steps: (1) PNA probe immobilization onto the transducer surface, (2) hybridization to target DNA strands, (3) indicator binding, and (4) chronopotentiometric transduction. The immobilized PNA probes showed outstanding order specificity and gave rise to rapid hybridization with the target oligonucleotides sequences (Ghosh and Mukhopadhyay 2013).

7.10 Immobilization in Chromatography

The basic principle of immobilization is applicable in affinity chromatography, a purification method based on interaction between ligand and receptor. Ligands are molecules that are immobilized on solid surface, whereas receptors are dissolved in mobile phase. Ligands can be carbohydrate based or protein based, which are immobilized on stable, inert, and macroporous matrix such as agarose, cellulose, silica, polyacrylamide, dextrose, and polystyrene (Armanda et al. 2015). Interaction between two molecules such as between enzyme and substrate, receptor and ligand, or antigen and antibody is the principle behind separation of biochemical mixtures. These interactions are specific, reversible, and biphasic, that is, ligand acts as stationary and receptor acts as mobile phase. The high selectivity of affinity chromatography is based on the specific interaction of desired molecule with the stationary phase due to which it remains trapped within the column, whereas undesired material that does not interact will elute first leading to the separation (Urh et al. 2009).

A novel technique utilizing immobilized artificial membrane chromatography was studied by Tsopelas et al. (2017) to predict the bioconcentration of pharmaceutical compounds in aquatic organisms. Nagami et al. (2014) developed metal affinity-immobilized liposome chromatography (MA-ILC) technique to separate peptides. Liposomes were immobilized along with functional ligand, N-hexadecyl iminodiacetic acid, that adsorbs metal ions. First, copper ion was adsorbed in gel matrix through its complex reaction with functional ligand and then retention of synthetic peptides was evaluated.

Based on affinity chromatography, immobilized lectins are used for separation and purification of glycoconjugates. As lectin and glycoconjugate interaction is specific, noncovalent, and reversible, lectins can be used as specific ligands in the development of affinity columns. In a research by Rêgo et al. (2016), PVA and GA were used as a support for immobilization of Concanavalin A, type of glucose-binding lectin. PVA-GA interpenetrated network was observed to be efficient for lectin covalent immobilization and as affinity chromatography matrix for entrapment of *Parkia pendula* gum.

References

Agyei, D., B. Shanbhag, and L. He. 2015. Enzyme engineering (immobilization) for food applications. In *Improving and Tailoring Enzymes for Food Quality and Functionality* (Ed.) R. Y. Yada, pp. 213–235. Tokyo, Japan: Woodhead Publishing.

Alloue, W. A., J. Destain, T. El Medjoub, H. Ghalfi, P. Kabran, and P. Thonart. 2008. Comparison of *Yarrowia lipolytica* lipase immobilization yield of entrapment, adsorption, and covalent bond techniques. *Applied Biochemistry Biotechnology* 150:51–63.

Anal, A. K. 2008. Controlled release dosage forms. In *Pharmaceutical Manufacturing Handbook* (Ed.) S. C. Gad, pp. 347–392. Hoboken, NJ: John Wiley & Sons.

Anal, A. K. and H. Singh. 2007. Recent advances in microencapsulation technologies for probiotics for industrial applications. *Trends in Food Science and Technology* 18:240–251.

Anal, A. K. and W. F. Stevens. 2005. Chitosan-alginate multilayer beads for controlled release of ampicillin. *International Journal of Pharmaceutics* 290:45–54.

Anal, A. K., D. Bhopatkar, S. Tokura, H. Tamura H, and W. F. Stevens. 2003. Chitosan-alginate multilayer beads for gastric passage and controlled intestinal release of protein. *Drug Development and Industrial Pharmacy* 9:713–724.

Annison, G., N. W. H. Cheetham, and I. Couperwhite. 1983. Determination of the uronic acid composition of alginates by high-performance liquid chromatography. *Journal of Chromatography* 204:137–143.

Ansari, S. A. and Q. Hussain. 2012. Potential applications of enzymes immobilized on/in nano materials: A review. *Biotechnology Advances* 30:512–523.

Armanda, D. H., J. T. Santos, M. R. Richards, and C. W. Cairo. 2015. Protecting group-free immobilization of glycans for affinity chromatography using glycosylsulfonohydrazide donors. *Carbohydrate Research* 417:109–116.

Atkins, E. D., I. A. Nieduszynski, W. Mackie, K. D. Parker, and E. E. Smolko. 1973. Structural components of alginic acid. 1. Crystalline structurs of poly-β-D-mannuronic acid. Results of X-ray diffraction and polarized infrared studies. *Biopolymers* 12:1865–1878.

Bagal, D. and M. S. Karve. 2006. Entrapment of plant invertase within novel composite of agarose-guar gum biopolymer membrane. *Analytica Chimica Acta* 555:316–321.

Beck, J., R. Angus, D. Madsen, D. Britt, B. Vernon, and K. T. Nguyen. 2002. Islet encapsulation: Strategies to enhance islet cell functions. *Tissue Engineering* 2:503–511.

Bhopatkar, D., A. K. Anal, and W. F. Stevens. 2005. Ionotropic alginate beads for controlled intestinal protein delivery: Effect of chitosan and barium counterions on entrapment and release. *Journal of Microencapsulation* 22:91–100.

Biesalski, H. K., L. O. Dragsted, I. Elmadfa et al. 2009. Bioactive compounds: Definition and assessment of activity. *Nutrition* 25:1202–1205.

Bunger, C. M., C. Gerlach, T. Freier et al. 2003. Biocompatibility and surface structure of chemically modified immunoisolating alginate-PLL capsules. *Journal of Biomedical Materials Research*. Part A. 67A:1219–1227.

Calafiore, R., G. Basta, G. Luca et al. 2006. Standard technical procedures for microencapsulation of human iselts for graft into nonimmunosuppressed patients with type 1 diabetes mellitus. *Transplant Proceedings* 38:1156–1157.

Carpentier, B. and O. Cerf. 1993. Biofilms and their consequences, with particular reference to hygiene in the food industry. *Journal of Applied Bacteriology* 75:499–511.

Chang, T. M. S. 1964. Semipermeable microcapsules. *Science* 146:524.

Chick, W. L., A. A. Like, and V. Lauris. 1975. Beta cell culture on synthetic capillaries: An artificial endocrine pancreas. *Science* 187:847–849.

Clayton, H. A., N. J. London, P. S. Colloby, P. R. Bell, and R. F. James. 1991. The effect of capsule composition on the biocompatibility of alginate-poly-l-lysine capsules. *Journal of Microencapsulation* 8:221–233.

Cole, D. R., M. Waterfall, M. Mclntyre, and J. D. Baird. 1992. Microencapsulated islet grafts in the BB/E rat: A possible role for cytokines in graft failure. *Diabetologia* 35:231–237.

Cunhan, A. G., G. Fernandez-Lorente, J. V. Bevilaqua et al. 2008. Immobilization of *Yarrowia lipolytica* lipase-a comparison of stability of physical adsorption and covalent attachment techniques. *Applied Biochemistry Biotechnology* 146:49–56.

Darrabie, M. D., W. F. Kendall, and E. C. Opara. 2005. Characteristics of poly-l-ornithine-coated alginate microcapsules. *Biomaterials* 26:6846–6852.

Datta, S., L. R. Christena, and Y. R. S. Rajaram. 2013. Enzyme immobilization: an overview on techniques and support materials. *3 Biotechnollgy* 3:1–9.

de Vos, P, C. G. van Hoogmoed, and H. J. Busscher. 2002. Chemistry and biocompatibility of alginate-PLL capsules for immunoprotection of mammalian cells. *Journal of Biomedical Materials Research Part A* 60:252–259.

de Vos, P. and P. Marchetti. 2002. Encapsulation of pancreatic islets for transplantation in diabetes: The untouchable islets. *Trends in Molecular Medicine* 8:363–366.

de Vos, P., B. J. De Haan, G. H. Wolters, J. H. Strubbe, and R. Van Schilfgaarde. 1997. Improved biocompatibility but limited graft survival after purification of alginate for microencapsulation of pancreatic islets. *Diabetologia* 40:262–270.

de Vos, P., C. G. van Hoogmoed, J. van Zanten, S. Netter, J. H. Strubbe, and H. J. Busscher. 2003. Long-term biocompatibility, chemistry and function of microencapsulated pancreatic islets. *Biomaterials* 4:305–312.

de Vos, P., G. H. J. Wolters, and R. van Schilfgaarde. 1996. Possible relationship between fibrotic overgrowth of alginate-polylysine-alginate microcapsulated rat islet allografts. *Transplantation* 62:893–899.

Desai, T. A., D. J. Hansford, and M. Ferrari. 2000. Micromachined interfaces: New approaches in cell immunoisolation and biomolecular separation. *Biomolecular Engineering* 17:23–36.

Dionne, K. E., B. M. Cain, R. H. Li et al. 1996. Transport characterization of membranes for immunoisolation. *Biomaterials* 17:257–266.

Duvivier-Kali, V. F., A. Omer, R. J. Parent, J. J O'Neil, and G. C. Weir. 2001. Complete protection of islets against allorejection and autoimmunity by a simple barium-alginate membrane. *Diabetes* 50:1698–1705.

Dwevedi, A. 2016. *Enzyme Immobilization Advances in Industry, Agriculture, Medicine and the Environment*. New York: Springer.

Elliot, R. B., L. Escobar, P. L. Tan et al. 2005a. Intraperitoneal alginate-encapsulated neonatal porcine islets in a placebo-controlled study with 16 diabetic cynomolgus primates. *Transplant Proceedings* 37:3505–3508.

Elliot, R. B., L. Escobar, R. Calafiore et al. 2005b. Transplantation of micro- and macroencapsulated piglet islets into mice and monkeys. *Transplant Proceedings* 37:468–469.

Elliott, R. B., L. Escobar, P. L. Tan, M. Muzina, S. Zwine, and C. Buchanan. 2007. Live encapsulated porcine islets from a type 1 diabetic patient 9.5 year after xenotransplantation. *Xenotransplantation*. 14:157–161.

Elnashar, M. M. M. 2010. Review article: Immobilized molecules using biomaterials and nanobiotechnology. *Journal of Biomaterials and Nanobiotechnology* 1:61–77.

Elnashar, M. M., E. N. Danial, and G. E. Awad. 2009. Novel carrier of grafted alginate for covalent immobilization of inulinase. *Industrial and Engineering Chemistry Research* 48:9781–9785.

Erogulu, E., S. M. Smith, and C. L. Raston. 2015. Application of various immobilization techniques for algal bioprocess. In *Biomass and Biofuels from Microalgae* (Eds.) N. R. Moheimani, M. P. McHenry, K. de Boer, and P. A. Bahri, pp. 19–44. New York: Springer.

Evstatieva, Y., M. Yordanova, G. Chernev, Y. Ruseva, and D. Nikolova. 2014. Sol–gel immobilization as a suitable technique for enhancement of α-amylase activity of *Aspergillus oryzae* PP. *Biotechnology and Biotechnological Equipment* 28:728–732.

Fang, Y., B. Kennedy, T. Rivera et al. 2012. Encapsulation system for protection of probiotics during processing. Google Patents.

Foresti, M. L. and M. L. Ferreira. 2007. Chitosan-immobilized lipases for the catalysis of fatty acid esterification. *Enzyme Microbiology Technology* 40:769–777.

Ghosh, S. and R. J. Mukhopadhyay. 2013. Nucleic acid sensing onto peptide nucleic acid (PNA) modified solid surfaces. *Journal of Bioanalysis and Biomedicine* 5:4.

Granicka, L. H., J. W. Kawaik, L. Glowacka, A. Wengski. 1996. Encapsulation of OKT3 cells in hollow fibers. *ASAIO Journal* 42:M863–M866.

Grohn, P., G. Klock, J. Schmitt et al. 1994. Large scale production of barium-alginate-coated islets of Langerhans for immunoisolation. *Experimental and Clinical Endocrinology* 102:380–387.

Hasse, C., A. Zeilke, G. Klock et al. 1998. Amitogenic alginates: Key to first clinical application of microencapsulation technology. *World Journal of Surgery* 22:659–665.

Horne, I., T. D. Sutherland, R. L. Harcourt, R. J. Russell, and J. G. Oakeshott. 2002. Identification of an (Organophosphate degradation) gene in an agrobactericum isolate. *Applied and Environmental Microbiology* 5:51.

Hu, X., L. Mu, J. Wen, and Q. Zhou. 2012. Immobilized smart RNA on graphene oxide nanosheets to specifically recognize and adsorb trace peptide toxins in drinking water. *Journal of Hazardous Materials* 30:387–392.

Idris, A. and A. Bukhari. 2011. Immobilized *Candida antarctica* lipase B: Hydration, stripping off and application in ring opening polyester synthesis. *Biotechnology Advances* 30:550–563.

Karel, S. F., S. B. Libicki, and C. R. Robertson. 1985. The immobilization of whole cells: Engineering principles. *Chemical Engineering Science* 40:1321–1354.

Kennedy, J. F., B. Kalogerakis, and J. M. S. Cabral. 1984. Surface immobilization and entrapping of enzymes on glutaraldehyde crosslinked gelatin particles. *Enzyme and Microbial Technology* 6:127–1231.

Kierek-Pearson, K. and E. Karatan. 2005. Biofilm development in bacteria. *Advances in Applied Microbiology* 57:79–111.

King, G. A., A. J. Daugulis, P. Faulkner, and M. F. A. Goosen. 1987. Alginate-polylysine microcapsules of controlled membrane molecular weight cut-off for mammalian cell culture engineering. *Biotechnology Progress* 3:231–240.

Kizilel, S., M. Garfinkel, and E. Opara. 2005. The bioartificial pancreas: Progress and challenges. *Diabetes Technology and Therapeutics* 7:968–985.

Klöck, G, H. Frank H, R. Houben et al. 1994. Production of purified alginates suitable for use in immunoisolated transplantation. *Applied Microbiology and Biotechnology* 40:638–643.

Klöck, G., A. Pfeffermann, C.Ryser et al. 1997. Biocompatibility of mannuronic acid-rich alginates. *Biomaterials* 18:707–713.

Kluseng, B, G. Skjak-Break, L. Ryan et al. 1999. Transplantation of alginate microcapsules: Generation of antibodies against alginates and encapsulated porcine islet-like-cell clusters. *Transplanatation* 67:978–984.

Koch, S., C. Scwinger, J. Kressler, C.H. Heinzen, and N. G. Rainov. 2003. Alginate encapsulation of genetically engineered mammalian cells: Comparison of production devices, methods and microcapsule characteristics. *Journal of Microencapsulation* 20:303–316.

Kris-Etherton, P.M., K. D. Hecker, A. Bonanome et al. 2002. Bioactive compounds in foods: Their role in the prevention of cardiovascular disease and cancer. *The American Journal of Medicine* 113:71–88.

Krol, S., S. del Guerra, M. Grupillo, A. Diaspro, A. Gliozzi, and P. Marchetti. 2006. Multilayer nanoencapsulation: New approach for immune protection of human pancreatic islets. *Nano Letters* 6:5828–5835.

Lanza, R. P., D. H. Butler, K. M. Borland et al. 1991. Xenotransplantation of canine, bovine and porcine islets in diabetic rats without immunosuppression. *Proceedings of the National Academy of Sciences of the United States of America* 88:11100–11104.

Lanza, R. P., R. Jackson, A. Sullivan et al. 1999. Xenotransplantation of cells using biodegradable microcapsules. *Transplantation* 67:1105–1111.

Lee, C. H., T. S. Lin, and C. Y. Mou. 2009a. Mesoporous materials for encapsulating enzymes. *Nano Today* 4:165–179.

Lee, J. I., R. Nishimura, H. Sakai, N. Sakai, and T. Kenmochi. 2008. A newly developed immunoisolated bioartificial pancreas with cell sheet engineering. *Cell Transplant* 17:51–59.

Lee, S. H., E. Hao, A. Y. Savinov, I. Geron, A. Y. Strngin, and P. Itkin-Ansari. 2009b. Human beta-cell precursors mature into functional insulin-producing cells in an immunoisolation device: Implications for diabetes cell therapies. *Transplantation* 87:983–991.

Li, R. H. 1998. Materials for immunoisolated cell transplantation. *Advanced Drug Delivery Reviews* 33:87–109.

Li, R. H., D. H. Altreuter, and F. T. Gentile. 1996. Transport characterization of hydrogel matrices for cell encapsulation. *Biotechnology and Bioengineering* 50:365–373.

Lim, F. and A. M. Sun. 1980. Microencapsulated islets as bioartificial endocrine pancreas. *Science* 210:908–910.

Liu, X., L. Luo, Y. Ding, Y. Xu, and F. Li. 2011. Hydrogen peroxide biosensor based on the immobilization of horseradish peroxidase on γ-Al_2O_3 nanoparticles/chitosan film-modified electrode. *Journal of Solid State Electrochemistry* 15:447–453.

LÖhr, J. M., R. Salier, B. Salmons, and W. H.Günzburg. 2006. Microencapsulation of genetically engineered cells for cancer therapy. *Methods Enzymology* 346:603–618.

Lum, Z. P., I. T. Tai, M. Krestow, I. Vacek, and I. M. Sun. 1992. An evaluation of new smaller capsules. *Transplantation* 53:1180–1183.

Lum, Z. P., I. T. Tai, M. Krestow, J. Norton, I. Vacek, and I. M. Sun. 1991. Prolonged reversal of diabetic state in NOD mice by microencapsulated rat islets. *Diabetes* 40:1511–1516.

Mazur, F., M. Bally, B. Städler, and R. Chandrawati. 2017. Liposomes and lipid bilayers in biosensors. *Advances in Colloid and Interface Science* 249:88–99. In Press.

McNaught, A. D. and A. Wilkinson. 1997. IUPAC. *Compendium of Chemical Terminology*. Oxford, UK: Blackwell Scientific Publications.

Mohamad, N. R., N. H. C. Marzuki, N. A. Buang, F. Huyop, F., and R. A. Wahab. 2015. An overview of technologies for immobilization of enzymes and surface analysis techniques for immobilized enzymes. *Biotechnology, Biotechnological Equipment* 29:205–220.

Mulagalapalli, S., S. Kumar, R. C. Kalathur, and A. M. Kayastha. 2007. Immobilization of urease from pigeonpea (*Cajanus cajan*) on agar tablets and its application in urea assay. *Applied Biochemistry Biotechnology* 142:291–297.

Muschler, G.F., C. Nakamoto, and L. G. Griffith LG. 2004. Engineering principles of clinical cell-based tissue engineering. *The Journal of Bone and Joint Surgery. American Volume* 86:1541–1558.

Nafea, E. H., A. Marson, L. A. Poole-Warre, and P. J. Martens. 2011. Immunoisolating semi-permeable membranes for cell encapsulation: Focus on hydrogels. *Journal of Controlled Release: Official Journal of the Controlled Release Society* 154:110–122.

Nagami, H., H. Umakoshi, T. Kitaura, G. L. Thompson III, T. Shimanouchi, and R. Kuboi. 2014. Development of metal affinity-immobilized liposome chromatography and its basic characteristics. *Biochemical Engineering Journal* 84:66–73.

Nakarani, M. and A. M. Kayastha. 2007. Kinetics and diffusion studies in urease-alginate biocatalyst beads. *Oriental Pharmacy Experimental Medicine* 7:79–84.

Narang, A.S. and R. I. Mahato. 2006. Biological and biomaterial approaches for improved islet transplantation. *Pharmacological Review* 58:194–243.

Nazari, T., M. Alijanianzadeh, A. Molaeirad, and M Khayati. 2016. Immobilization of subtilisin carlsberg on modified silica gel by cross-linking and covalent binding methods. *Biomacromolecular Journal* 2:53–58.

O'Neill, G. J., T. Egan, J. C. Jacquier, M. O'Sullivan, and E. Dolores O'Riordan. 2016. Whey microbeads as a matrix for the encapsulation and immobilization of riboflavin and peptides. *Food Chemistry* 160:46–52.

O'Shea, G. M. and A. M. Sun. 1986. Encapsulation of rat islets of Langerhans prolongs xenograft survival in diabetic mice. *Diabetes* 35:943–946.

Opara, E. C. and W. F. Jr. Kendal. 2002. Immunoisolation techniques for islet cell transplantation. *Expert Opinion on Biological Therapy* 2:503–511.

Orive, G., A. M. Carcaboso, R. M. Hernandez, A. R. Gascon, and J. L. Pedraz. 2005. Biocompatibility evaluation of different alginates and alginate-based microcapsules. *Biomacromolecules* 6:1321–1331.

Orive, G., S. K. Tam, J. L. Pedraz, and J. P. Halle. 2006. Biocompatibility of alginate poly-l-lysine microcapsules for cell therapy. *Biomaterials* 27:3692–3700.

Portero, A., G. Orive, R. M. Hernandez, and J. L. Pedraz. 2010. Cell encapsulation for the treatment of central nervous system disorders. *Revista de Neurologia* 50:409–419.

Pundir, C. S. 2015. *Enzyme Nanoparticles*. Tokyo, Japan:William Andrew.

Qi, M., B. L. Strand, Y. Mφrch et al. 2008. Encapsulation of human islets in novel inhomogenous alginate-Ca2+/ba2+ microbeads: *In vitro* and *in vivo* function. *Artificial Cells, Blood Substitutes, and Immobilization Biotechnology*. 36:403–420.

Rêgo, M. J. B. de M., L. R. A. de Lima, A. F. P. Longo, G. A. Nascimento-Filho, L. B. de Carvalho Júnior, and E. I. C. Beltrão. 2016. PVA-Glutaraldehyde as support for lectin immobilization and affinity chromatography. *Acta Sclentlarum*. 38:291–295.

Rihová, B. 2000. Immunocompatibility and biocompatibility of cell delivery systems. *Advanced Drug Delivery Review* 42:65–80.

Sakai, S., T. Ono, H. Ijima, and K. Kawakami. 2001. Synthesis and transport characterization of alginate/aminopropyl-silicate/alginate microcapsule: Application to bioartificial pancreas. *Biomaterials* 22:2827–2834.

Siebers, U., T. Zekon, A. Horcher et al. 1992. *In vitro* testing of rat and porcine islets microencapsulated in barium alginate beads. *Transplant Proceedings* 24:950–951.

Skinner, S. J. M., M. S. Geaney, H. Lin et al. 2009. Encapsulated living choroid plexus cells: Potential long-term treatments for central nervous system disease and trauma. *Journal of Neural Engineering*. 6:1–11.

Skinner, S. J. M., M. S. Geaney, R. Rush, et al. 2006. Choroid plexus transplants in the treatment of brain diseases. *Xenotransplantation* 13:284–288.

Skjak-Break, G., E. Murano, and S. Paoletti. 1989. Alginate as immobilization material II: Determination of polyphenol contaminants by fluorescence spectroscopy, and evaluation of methods for their removal. *Biotechnology and Bioengineering* 33:90–94.

Sona, P. 2010. Nanoparticulate drug delivery systems for the treatment of diabetes. *Digest Journal of Nanomaterials and Biostructures* 5:411–418.

Soon-Shiong, P., E. Feldman, R. Nelson et al. 1992. Long-term reversal of diabetes in the large animal model as encapsulated islet transplantation. *Transplant Proceedings* 246:2946–2947.

Standard guide for characterization and testing of alginates as starting materials intended for use in biomedical and tissue-engineered medical products applications. 2006; ASTM International Designation F2064-00.

Tam, S. K., B. J. de Haan, M.M. Faas, J. P. Hallé, L. Yahia, and P. de Vos. 2009. Adsorption of human immunoglobulin to implantable alginate-poly-L-lysine microcapsules: Effect of microcapsule composition. *Journal of Biomedical Materials Research. Part A*. 89:609–615.

Tam, S. K., J. Dusseault, S. Polizu, M. Menard, J. P. Halle, and L. Yahia. 2006. Impact of residual contamination on the biofunctional properties of purified alginates used for cell encapsulation. *Biomaterials* 27:1296–1305.

Tam, S. K., S. Bilodeau, J. Dusseault, G. Langlois, J.-P. Hallé, and L. H. Yahia. 2011. Biocompatibility and physicochemical characteristics of alginate-polycation microcapsules. *Acta Biomaterialia* 7:1683–1692.

Taunk, A., K. K. K. Ho, G. Iskander, M. D. P. Willcox, and N. Kumar. 2016. Surface immobilization of antibacterial quorum sensing inhibitors by photochemical activation. *Journal of Biotechnology and Biomaterials* 6:3.

Teramura, Y. and H. Iwata. 2010. Bioartificial pancreas: Microencapsulation and conformal coating of islet of Langerhans. *Advanced Drug Delivery Reviews* 62:827–840.

Terpou, A., A. Bektorou, M. Kanellaki, A. A. Koutinas, and P. Nigam. 2017. Enhanced probiotic viability and aromatic profile of yogurts produced using wheat bran (*Triticum aestivum*) as cell immobilization carrier. *Process Biochemistry* 55:1–10.

Thanos, C. G., B. E. Bintz, W. J. Bell et al. 2006. Intraperitoneal stability of alginate-polyornithine microcapsules in rats: An FTIR and SEM analysis. *Biomaterials* 27:3570–3579.

Thu, B., P. Bruheim, T. Espevik, O. Smidord, P. Soon-Shiong P, and G. Skjak-Break. 1996. Alginate polycation microcapsules. I. Interaction between alginate and polycations. *Biomaterials* 17:1031–1040.

Tosa, T., T. Mori, N. Fuse, and I. Chibata. 1966. Studies on continuous enzyme reactions. I. Screening of carriers for preparation of water-insoluble aminoacylase. Enzymologia 31:214–224.

Tseng, T. T. C., C. F. Chang, and W. C. Chan. 2014. Fabrication of implantable, enzyme-immobilized glutamate sensors for the monitoring of glutamate concentration changes *in vitro* and *in vivo*. *Molecules* 19:7341–7355.

Tsopelas, F., C. Stergiopoulour, L. A. Tsakanika, M. Ochsenkuhn-Petropoulou, and A. Tsantili-Kakoulidou. 2017. The use of immobilized artificial membrane chromatography to predict bioconcentration of pharmaceutical compounds. *Ecotoxicology and Environmental Safety* 139:150–157.

Tziampazis, E. and A. Sambanis. 1995. Tissue engineering of a bioartificial pancreas: Modeling the cell environment and device function. *Biotechnology Progress* 11:115–126.

Urh, M., D. Simpson, and K. Zhao. 2009. Affinity chromatography. In *Methods in Enzymology Guide to Protein Purification*, (Eds.) R. R. Burgess and M. P. Deutscher, pp. 417–438. Tokyo, Japan: Elsevier.

Visted, T., R. Bjerkvig, and P. O. Enger. 2001. Cell encapsulation technology as a therapeutic strategy for CNS malignancies. *Neuro-oncology* 3:201–210.

Wang, N., G. Adams, L. Buttery, F. H. Falcone, and S. Stolink. 2009. Alginate encapsulation technology supports embryonic stem cells differentiation into insulin-producing cells. *Journal of Biotechnology* 144:304–312.

Wang, T., I. Lacík, M. Brissová et al. 1997. The encapsulation system for the immuno-isolation of pancreatic islets. *Nature Biotechnology* 15:358–362.

Wen, H., V. Nallathambi, D. Chakraborty, and S. C. 2011. Barton. Carbon fiber micro-electrodes modified with carbon nanotubes as a new support for immobiliza-tion of glucose oxidase. *Microchimica Acta* 175:283–289.

WikstrÖm, J., M. Elomma, H. Syväjävi et al. 2008. Alginate-based microencapsulation of retinal pigment epithelial cell line for cell therapy. *Biomaterials* 29:869–876.

Wilson, J. T. and E. L. Chaikof. 2008. Challenges and emerging technologies in the immunoisolation of cells and tissues. *Advanced Drug Delivery Review* 60:124–145.

Wilson, J. T., W. Cui, and E. L. Chaikof. 2008. Layer-by-layer assembly of a confor-mal nanothin PEG coating for intraportal islet transplantation. *Nano Letters* 8:1940–1948.

Wollert, K. C. and H. Drexler. 2006. Cell-based therapy for heart failure. *Current Opinion in Cardiology* 21:234–239.

Zhang, X. and V. K. Yadavalli. 2011. Surface immobilization of DNA aptamers for biosensing and protein interaction analysis. *Biosensors and Bioelectronics* 26:3142–147.

Zhang, X., H. He, C. Yen, W. Ho, and L. J. Lee. 2008. A biodegradable, immuno-protective, dual nanoporous capsule for cell-based therapies. *Biomaterials* 29:4253–4259.

Zhang, Y. Q., M. L. Tao, W. D. Shen et al. 2004. Immobilization of L-asparaginase on the microparticles of the natural silk sericin protein and its characters. *Biomaterials* 25:3751–3759.

Zhao, X., F. Qi, C. Yuan, W. Du, and D. Liu. 2015. Lipase-catalyzed process for bio-diesel production: Enzyme immobilization process simulation optimization. *Renewable and Sustainable Energy Reviews* 44:182–197.

Zhao, Z., X. Xie, Z. Wang et al. 2016. Immobilization of *Lactobacillus rhamnosus* in mes-oporous silica-based material: An efficiency continuous cell-recycle fermenta-tion system for lactic acid production. *Journal of Bioscience and Bioengineering* 121:645–651.

Zhou Y., T. Sun, M. Chan et al. 2005. Scalable encapsulation of hepatocytes by electro-static spraying. *Journal of Biotechnology* 17:99–109.

Zimmermann, Y. S., P. Shahgaldian, P. F. X. Corvini, and G. Hommes. 2011. Sorption-assisted surface conjugation: A way to stabilize laccase enzyme. *Applied Microbiology and Biotechnology* 92:169–178.

8

Nanostructure-Based Delivery Dosage Forms in Pharmaceuticals, Food, and Cosmetics

8.1 Introduction

The advancement in the field of nanoscience and nanotechnology holds the great potential to revolutionize the scope in various fields. Food nanotechnology is an emerging area with great scope to generate innovative products and processes. Nanotechnology refers to the design, characterization, production and application of structures, devices, and systems by controlling the shape and size at nanoscale, that is, magnitude of 10^{-9} m. The systems at nanoscale size range have the benefits of high surface to volume ratio, which seems to be beneficial in innovation and development in the field of medical diagnostic, therapeutics, energy, molecular computing, and structural materials. In food technology, nanotechnology has transformed the entire system from production of crops to processing, storage, and development of innovative products. The major applications of these techniques involve nanoencapsulation for controlled delivery of nutraceuticals, nanocomposite in packaging materials, and nanobiosensors for detecting microbial and physical contamination (Anandharamakrishnan 2014).

8.2 Nanoencapsulation

Encapsulation is the process of incorporation of active metabolites either in solid, liquid, or gaseous state in small capsules, which may range from nanometer to millimeter in size. Encapsulation aims to serve various purposes such as to protect the active sensitive compounds by reducing its exposure to the surrounding (light, air, heat, and moisture), promote easy handling, and assist in controlled release attributes. This technique dates to 1950s, when dyes were first encapsulated in the gel matrix of gelatin and gum arabic, produced by Green and coworkers in the development of carbonless copying paper. Encapsulation technology since then has been employed in different industries such as food, pharmaceuticals, cosmetics, chemicals, printing, and so on (Mishra 2016).

Microencapsulation is the process of entrapping core materials (solid, liquid, or gas form) in microcapsules with the diameter ranging from few microns to 1 mm. For encapsulation, semipermeable, thin and strong membranes are used, which are made from food-grade polymers such as chitosan, carrageenan, alginate, carboxymethyl cellulose, gelatin, and pectin. In food industry, microencapsulation served to stabilize core material, control oxidative reaction, provide both temporal and time-controlled release, mask undesirable flavor and odor, reduce the nutritional loss, and extend the shelf life (Anal and Singh 2007).

Nanoencapsulation is an area with a rapid exposure because of its potential applications in the food, pharmaceutical, and cosmetics industry. This method has been utilized for the protection of the bioactive compounds such as polyphenols, enzymes, antioxidants, micronutrients, and nutraceuticals. Nanoencapsulation is a process of encapsulating bioactive compounds in miniature form at the nanoscale range. As particle size directly affects in the delivery of bioactive compounds, that is, reduced size prolongs gastrointestinal retention time and adhesiveness to epithelial; hence, bioactive compounds encapsulated in nanocapsules or nanoparticles with diameter in range of 10–1000 nm tend to have greater potential in target delivery compared to microcapsules with diameter in the range of 3–800 μm. Vesicular systems with a confined bioactive compounds to a cavity, which is surrounded by a unique membrane of polymer are known as nanocapsules, and the nanospheres are the matrix systems where bioactive compound can be found in a uniform dispersion (Ezhilarasi et al. 2013). The schematic representation of nanocapsules and nanospheres encapsulating bioactive compounds is shown in Figure 8.1.

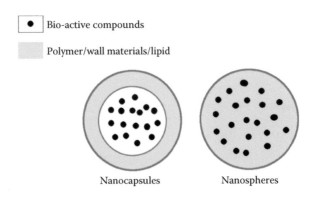

Nanocapsules   Nanospheres

FIGURE 8.1
Schematic structure of nanocapsules and nanospheres. (From Ezhilarasi, P.N. et al., *Food Bioproc. Technol.*, 6, 628–647, 2013. With Permission.)

8.3 Materials Used for Nanoencapsulation

As encapsulation is the process of casing active compound in the matrix of protecting membrane structures; there are two main components: (1) encapsulants, which refer to the core materials that need to be encapsulated and (2) encapsulating agents that are used as wall-forming materials. There are varieties of wall materials available to be used as encapsulating agents and they are as follows:

1. *Waxes and lipids*: Beeswax, wax micro- and macroemulsion, glycerol distearate, and natural and modified fats
2. *Proteins*: Whey protein, soy protein isolates, gelatins, gluten, zein, and others
3. *Carbohydrates*: Cellulose acetate, alginates, carrageenan, chitosan, and so on
4. *Food-grade polymers*: Polypropylene, polyvinyl acetate, and polystyrene, polybutadiene

The materials that are usually encapsulated include flavors, antimicrobial agents, nutraceuticals, vitamins, minerals, antioxidants, enzymes, and so on (Lakkis 2016). On the basis of the functional group, biomacromolecules may be either polycationic or polyanionic in nature. Proteins and polysaccharides are widely utilized biopolymers in the food, cosmetic, and pharmaceutical industries. On account of their amphiphilic nature, proteins are widely utilized as emulsifying agents and polysaccharides such as chitosan, alginate, and resistant starch acts as emulsifying, stabilizing, and thickening agents (Anal et al. 2008). Further, protein and polysaccharide conjugated together by Maillard reaction are found to be better encapsulating materials than the counterparts alone (Nasrin and Anal 2014, 2015). Various supplementary bioactive compounds or nutraceuticals encapsulated in nanocarrier form known as *nanoceutical*, have improved bioavailability, delivery, and solubility. These bioactive compounds can be hydrophilic or lipophilic based on their affinity to water. Common hydrophilic compounds that have been nanoencapsulated include polyphenols and ascorbic acids. Nanoencapsulated lipophilic compounds include lycopene, beta-carotene, lutein, phytosterol and docosahexaenoic acids (DHA) (Anandharamakrishnan 2014).

Lipid or natural polymer-based nanocarrier food systems are more often used in the encapsulation. Nanocochleates, archaeosomes, and nanoliposomes are nanocarrier systems, which are lipid based with tremendous applications in food, pharmaceutical, and cosmetic industries. Similarly, nanodelivery systems formulated using the natural polymers such as gelatin, alginate, collagen, chitosan, and α-lactalbumin are also widely used in various industries. Development of the food nanocarrier systems have been rapidly growing, which have led to improvement in bioavailability of nutraceuticals, development of nanodrops mucosal delivery system of vitamins and mineral delivery system (Ezhilarasi et al. 2013).

8.4 Nanoencapsulation Technique

Among the available different nanoencapsulation techniques, the best-suited method is selected on the basis of the physical and chemical properties of encapsulant and encapsulating agents and the intended application. The factors that need to be considered during the selection of encapsulation method include physicochemical properties and function of core material, appropriateness and feasibility of processing condition, amount of core material to be encapsulated, interaction between core and encapsulating agents, and protecting behavior and release properties of the encapsulating agents. Techniques that are used for the formation of the nanoencapsulation are more complicated than the techniques in microencapsulation. This complication is mainly because of the morphological behavior, core material, and releasing rate of the nanoencapsulant (Mishra 2016).

For the production of nanocapsules, different techniques such as emulsification, emulsification–solvent evaporation, coacervation, inclusion complexation, nanoprecipitation, and supercritical fluid techniques are useful. These techniques are either top–down approaches, which involve gradual size reduction with utilization of precise tools or bottom–down approaches that include self-organizing and assembling molecules influenced by pH, temperature, ionic strength, and the concentration. Emulsification and emulsification–solvent evaporation methods are classified under the top–down approach, whereas methods such as inclusion complex, nanoprecipitation, complexion, coacervation, and supercritical fluid techniques are examples of bottom–up approach. Various lipophilic and hydrophilic bioactive compounds can be encapsulated by using these nanoencapsulation techniques. Coacervation, emulsification, and supercritical fluid technique are used for both lipophilic and hydrophilic compounds. Lipophilic compounds are mostly used with the inclusion, emulsification, emulsification–solvent evaporation, and complexation methods (Ezhilarasi et al. 2013). Some of the nanoencapsulation methods are listed in Table 8.1.

TABLE 8.1

Examples of Nanoencapsulation Techniques of Various Bioactive Compounds

Encapsulation Method	Bioactive Compounds	Encapsulating Agents	References
Ionotropic gelation	Ampicillin	Chitosan–alginate	Anal and Stevens (2005), Anal et al. (2006)
Emulsification	β-carotene	Whey protein–dextran conjugate	Fan et al. (2017)
Emulsification–solvent evaporation	Lutein	Whey protein isolate	Teo et al. (2017)
Coacervation	Astaxanthin-containing lipid extract	Gelatin–cashew gum complex	Gomez-Estaca et al. (2016)
Inclusion complex	Omega-3 fatty acids	β-cyclodextrin complexes	Ünlüsayin et al. (2016)
Nanoprecipitation	Curcumin	Polyethylene glycol-*b*-poly(DL-lactide)	Chow et al. (2015)
Supercritical anti-solvent techniques	Lutein	Zein	Hu et al. (2012)

8.5 Nanoemulsion

An emulsion is the colloidal dispersion of immiscible solvents stabilized by emulsifiers. On the basis of phase distribution, emulsions can be classified as oil in water (o/w) emulsion where oil is dispersed in a continuous phase of water or water in oil (w/o) emulsion with water droplets dispersed in oil. Emulsion can also be multilayered such as oil in water in oil (o/w/o) emulsion or water in oil in water (w/o/w) (Anandharamakrishnan 2014).

Nanoemulsions, also known as miniemulsions, ultrafine emulsions, submicron emulsions, fine-dispersed emulsions, have different definition on the basis of the size range. Nanoemulsions are defined as nonequilibrium emulsions system with the droplet size in the range of 20–200 nm. They are also known as kinetically stable transparent system with mean diameter in the range of 100–500 nm or the thermodynamically stable isotropic system with droplet diameter in the range between 10 and 100 nm (Aboofazeli 2010). Basically, nanoemulsions are colloidal particulate with droplet size varying from 10 to 100 nm, comprising oil, water, and an emulsifier. Nanoemulsions have been developed with enhanced properties to act as carrier molecules for lipophilic bioactive compounds compared to conventional emulsions (Salvia-Trujillo et al. 2016).

Physicochemical properties of nanoemulsions such as high optical clarity, improved stability to gravitational separation and aggregation, and the ability to enhance the bioavailability of encapsulated active ingredients have increased interest toward the industrial application of nanoemulsions. On account of the small droplet size, nanoemulsions occupy a large surface area and can make strong interactions with biological components in the gastrointestinal tract (GIT). This signifies nanoemulsions have a high digestion rate in the GIT than the conventional emulsions because of more binding sites available for the digestive enzymes such as lipase. Furthermore, the smallest droplet size of nanoemulsions may enhance the transfer of naturally occurring hydrophobic bioactive compounds found in foods into the oil droplets (Salvia-Trujillo et al. 2013).

In past 20 years, different types of emulsion systems have been invented for encapsulation of various bioactive compounds and drugs because of their great potential to act as protective encapsulation system, high thermodynamic stability, and successful delivery. Thus, emulsion-based encapsulation system has been an essential approach to develop functional formulation in many applications, especially in food and pharmaceutical industries (Mishra 2016). Different types of emulsion serve different functions such as o/w emulsion have been used to encapsulate lipophilic components such as carotenoids, beta-carotene, plant sterols, and essential fatty acids, whereas w/o emulsion system has been used to encapsulate lipophilic compounds such as polyphenols (Ezhilarasi et al. 2013). Recently, nanoemulsions are widely used in the pharmaceutical industry because nanoemulsions are found to be more stable orally, on the skin and mucous membranes for drug delivery, and cosmetic industry.

8.6 Nanoemulsification Techniques

Nanoemulsions with uniform and small droplets are transparent or translucent in appearance, possess kinetic stability but are not thermodynamically stable. On account of its nonequilibrium nature, nanoemulsion cannot be developed spontaneously and requires energy input, either from mechanical devices or from the physicochemical properties of the components. Broadly, nanoemulsion fabrication can be done by two methods: (1) high-energy and (2) low-energy methods. Mechanical devices such as high-pressure homogenizer, microfluidizer, and sonicators have been used for high-energy emulsification methods and both o/w and w/o nanoemulsions have been developed from this method. On the other hand, low-energy emulsification method based on utilization of chemical properties of the system is achieved by a method such as phase-inversion method and self-emulsification technique (Solè and Solans 2013).

8.6.1 High-Energy Nanoemulsification

During a high-energy nanoemulsificaiton process, intense disruptive force is applied because of which the two opposite processes, that is, droplet coalescence and droplet disruption take place in the sample to be emulsified. The formation of smaller droplet occurs with the balance between these two processes (Anandharamakrishnan 2014). High-energy processes utilize particular machines such as high-pressure homogenizers, microfluidizers, and ultrasonic devices, which are capable of creating disruptive forces greater than the restorative forces holding the droplets into spherical shapes. The restorative forces, as determined by Laplace pressure: $\Delta P = \gamma/2r$, signifies that restorative force increases with increasing interfacial tension (γ) and decreasing droplet radius (r). The characteristics of nanoemulsions, which are prepared from high-energy method, are mainly affected by the operation conditions of mechanical devices such as intensity of energy used, processing time, temperature, nature, and concentration of the surfactant, and the physicochemical properties of the oil and water phases such as interfacial tension and viscosity (McClements 2010). In high-energy processes, the input energy density (ε) is 10^8–10^{10} W kg^{-1} (Gupta et al. 2016).

Macroemulsions are prepared simply by mixing oil, water, and surfactant for an adequate period of time as the beginning step of high-energy method. Then as the second step, the macroemulsions are converted into nanoemulsions by using a homogenizer where larger droplets are being pushed by a high-pressure pump to break into small droplets. The homogenization process has to repeat several times until the droplet size becomes constant (Mason et al. 2006). High-pressure homogenizers are commonly utilized in food industries to produce small droplet emulsions. Coarse emulsion produced from rotor–stator systems is passed in high-pressure homogenizer where the sample is pulled into a chamber by the backstroke of a pump and forced by forward stroke through a narrow value at the end of the chamber. On passage through the narrow valve, coarse emulsions are forced to combination of intense disruptive force, shear stress, cavitation, and turbulent flow, which causes the disruption of larger droplets into smaller. High-pressure homogenizer requires higher pressure and multiple passes through the system to produce the nanoemulsions. The working principle of microfluidizer is similar to the high-pressure homogenizer, in which high pressure is applied to convert coarse emulsion to fine nanoemulsion. (Anandharamakrishnan 2014).

Ultrasonic nanoemulsification with the application of ultrasound waves (20–100 MHz) is a relatively new approach in the sector of food industry. The basic working principle behind ultrasound is the cavitation process, that is, ultrasound waves cause the formation and collapse of microbubble at interface of continuous and dispersed phase under high-intensity acoustic field. Break down of bubble causes high velocity jets in liquid because of the production of intense shock waves and localized turbulence. Further, droplet

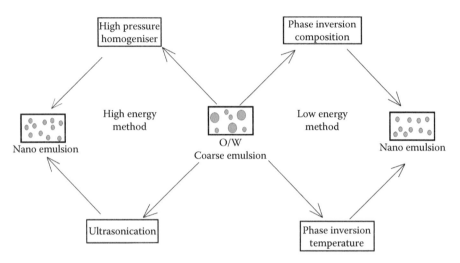

FIGURE 8.2
Representation of high-energy and low-energy nano-emulsification. (From Gupta, A. et al., *Soft Matt.*, 12, 2826–2841, 2016. With Permission.)

disruption is facilitated by intense shear forces produced from high-velocity jet (Abbas et al. 2013). The common high-energy and low-energy nanoemulsification methods are shown in Figure 8.2.

8.6.2 Low-Energy Nanoemulsification

In contrary, low-energy nanoemulsification processes are based on the spontaneous formation of emulsions and mainly affected by the phase behavior of certain surfactants, oil, and water systems. In this approach, inversion of the spontaneous curvature of the surfactant film from positive to negative or vice versa is uncertain. When there is inversion or change in spontaneous curvature of the surfactant film during emulsification, the process is termed as *phase inversion*. This process can be triggered by external factors. During phase-inversion temperature (PIT), phase inversion is assisted by temperature and in phase-inversion composition (PIC), change in composition leads to phase inversion. At a constant temperature, nanoemulsion formation may also occur without phase conversion and the process is known as *self-emulsification*. In case of self-emulsification, nanoemulsion formation is triggered by the rapid diffusion of surfactant and/or solvent molecules from the dispersed phase to the continuous phase. The occurrence of self-emulsification, in free surfactant system, is known as *Ouzo effect* (Solans and García-Celma 2017). The requirement of input energy density (ε) is very low (10^3–10^{15} W kg^{-1}) when compared with the high-energy methods. Therefore, a simple batch stirrer can be easily used to obtain the required energy input. The formation of smaller droplets in the low-energy method

TABLE 8.2

Examples of Nanoemulsions for Encapsulation of Different Bioactive Compounds

Techniques	Bioactive Compounds	Oil Phases	Emulsifiers	References
High-pressure homogenization	Astaxanthin	Soybean oil	Sodium caseinate	Khalid et al. (2017)
Ultrasound-assisted emulsification	Stinging nettle essential oil	Nettle oil	Surfactant T40	Gharibzahedi (2017)
Solvent displacement methods	β-carotene	High oleic sunflower oil	Tween 80	Oliveira et al. (2016)
Emulsification-evaporation methods	Lutein	Corn oil	Whey protein isolate	Teo et al. (2017)
Spontaneous emulsification	Cinnamon oil	Coconut oil	Tween 80 and sulfuric acid	Yildirim et al. (2017)

is mainly because of the phase inversion of the system. Low-energy process commences with a w/o macroemulsion, which is then converted into an o/w nanoemulsions, with a change in either composition or temperature. At the emulsion inversion point (EIP), w/o macroemulsions are prepared at room temperature followed by a slow dilution with water. The inversion point is that point where w/o emulsions are converted into o/w emulsion during the dilution process. At this point, small droplets are formed because of low oil–water interfacial tension value (Gupta et al. 2016). Different types of nano-emulsions utilizing variety of emulsifiers and oil phases for encapsulation of different types of bioactive compounds are listed in Table 8.2.

8.7 Nanoemulsions Finishing Techniques

The radius of nanoemulsions (usually below 100 nm) depends on the emulsion formation condition and the emulsifier type. Further, composition of oil droplets will determine the rate and extent of oil digestibility. Once the nanoemulsions are formed by low- or high-energy method, the droplet characteristics are further modified and the process is known as finishing techniques. Droplet characteristics such as interfacial properties, droplet size, droplet composition, and droplet physical state can be modified.

The change in interfacial properties may be useful to stabilize the emulsion. Interfacial properties such as composition, charge, rheology, or thickness can be changed by displacing the original emulsifier or by interfacial deposition method. Droplet sizes of nanoemulsions are further reduced by solvent displacement or evaporation methods. The composition of oil droplets is changed by mixing nanoemulsion with another emulsion or microemulsion system. Further, the physical state of droplets can be changed by full or partial crystallization of oil droplets in nanoemulsions (McClements 2010).

8.8 Nanoemulsions as Delivery Systems

The main objective of designing delivery systems is to regulate the digestion and to release and absorb the lipophilic components within the GIT to develop several types of applications in the food, pharmaceutical, and cosmetic industries. Basically, a delivery system can be used to transport and release the bioactive compounds and drugs at a targeted location within the GIT such as mouth, stomach, small intestine, or colon (Kosaraju 2005). Some of the emulsion systems utilized to deliver bioactive compounds are listed in Table 8.3.

There are several reasons that these compounds cannot be directly incorporated with foods. Few of those reasons could be low bioavailability or poor bioaccessibility, low water-solubility, poor chemical stability, and high melting point. Hence, it is vital to develop delivery systems that are able to overcome these problems to effectively transport bioactive compounds to the desired locations. Different delivery systems based on emulsion have been developed to encapsulate, protect, and deliver lipophilic bioactive compounds. Some of these emulsion-based structured delivery systems include conventional emulsion, nanoemulsion, multilayer emulsion, multiple emulsion, coated and filled

TABLE 8.3

Examples of Bioactive Components Delivery by Nanoemulsions System

Name	Types	Potential Nutritional Benefits	References
Fatty acids	Polyunsaturated fatty acids (EPA, DHA)	Regulate cardiovascular diseases, enhance immune response disorders, anticancer and bone health, stroke prevention	Mao and Miao (2015)
Carotenoids	β–carotene, lycopene, lutein, zeaxanthin, astaxanthin	Control cardiovascular diseases, anticancer, improve skin health, provide vitamin A	
Vitamins	Vitamin A, D, E	Antioxidant, improve visual, bone, skin, immune system, prevent cancer	
Phytosterols	Stigmasterols, β–sitosterols, Campesterols	Prevention of coronary heart diseases	McClements and Li (2010)
Nutraceuticals	Coenzyme Q	Control cardiovascular diseases, anticancer, antidiabetic, hypertension	

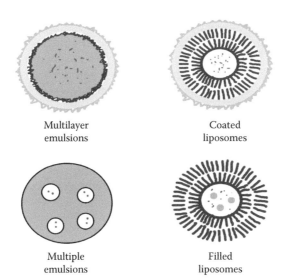

Multilayer emulsions

Coated liposomes

Multiple emulsions

Filled liposomes

FIGURE 8.3
Illustration of different emulsion-based delivery system. (From McClements, D.J., *J. Food Sci.*, 75, 30–42, 2010. With Permission.)

liposomes, coated and filled hydrogel particle, solid lipid particles, and colloidosomes (McClements 2010). Some of the common emulsion-based delivery systems are illustrated in Figure 8.3.

8.9 Nanoemulsions in Food Systems

The valuable bioactive compounds are difficult to incorporate directly into food matrix because of poor compatibility with food, susceptibility to degradation, and vulnerability to digestion with poor absorption profile. Therefore, bioactive compounds need to be encapsulated for easier fortification in food system and better stability and target. Moreover, the stability and the bioavailability of those compounds are highly affected by the nature of food component and their behavior in GIT. Therefore, it is important to enhance the bioavailability of bioactive compounds incorporated in food products and to achieve that several techniques have been developed, such as colloidal delivery systems. This system can be used to incorporate lipophilic compounds in different forms such as nanoemulsions, emulsion, or solid lipid nanoparticles. When compared with the conventional emulsions, nanoemulsions have shown better carrier properties to deliver lipophilic bioactive compounds as they contain more number of digestive enzyme binding sites (Salvia-Trujillo et al. 2016).

Bioavailability of bioactive compounds depends on bioaccessibility, absorption, and transformation of bioactive compounds in the GIT. The concentration of bioactive compounds that are ready to be absorbed in the GIT compared with the total ingested compounds is known as the bioaccessibility. Neutraceuticals in powdered forms are effectively transferred from bioactive crystalline form to the oil phase using nanoemulsions, which further regulate solubility in the GIT fluids. Solubility and bioaccessibility of curcumin molecules were reported to be increased in nanoemulsions. Curcumin molecules were transferred from curcumin to oil droplets in the emulsion during incubation at elevated temperatures, which increased the subsequent bioaccessibility of the curcumin when the emulsion was passed through simulated GIT conditions (Zou et al. 2015).

When considering the plant-based foods, the transportation of hydrophobic bioactive compounds from the plant tissue is facilitated by the lipid droplets in the nanoemulsion by acting as nonpolar solvent (Antoine et al. 2013). Supporting this statement, a study has found improved bioaccessibility of carotenoids in yellow pepper after being incorporated with nanoemulsions (Liu et al. 2015). The enhancement of bioaccessibility by nanoemulsions is mainly affected by the characteristics such as amount of lipid, size, and composition. It has been found that the bioaccessibility would significantly increase with the increasing amount of oil concentration (Salvia-Trujillo et al. 2013). Different nanoemulsion systems utilized to enhance bioavailability of the important bioactive compounds is listed in Table 8.4.

TABLE 8.4

Examples of Nanoemulsion Delivery System to Enhance Bioavailability of Bioactive Compounds

Emulsion Type	Bioactive Compounds	Factors Influencing Nutraceuticals Bioaccessibility	References
Corn oil emulsions	Curcumin	Emulsifier type and droplet size	Zou et al. (2015)
Corn oil emulsions	Curcumin	Oil type and concentration	Ahmed et al. (2012)
Corn oil, medium chain triglycerides or orange oil emulsions	Carotenoids in yellow peppers	Yellow peppers with corn oil nanoemulsions as excipient emulsions	Liu et al. (2015)
Olive oil emulsions	Carotenoids in carrot and tomato suspensions	Adding olive emulsions to carrot and tomato suspensions increased	Moelants et al. (2013)
Peanut oil emulsions	Carotenoids in tomato juice	Emulsification process and emulsifier type	Antoine et al. (2013)
Various emulsions	Carotenoids in vegetables and salads	Addition of oil (depending on fatty acid type) to salads and vegetables	Nagao et al. (2013)

8.10 Nanoemulsions in Pharmaceutical Industry

An effective drug delivery system will maximize the therapeutic effect of drugs while reducing the toxic effects. Generally, pills are used to deliver the required dosage of drugs, but with the development of science and technology, novel drug delivery systems have been introduced. Nanoemulsions are becoming more promising in the industry as effective delivery systems. Different types of nanoemulsions (w/o or o/w) with mean droplets size in range between 50 and 1000 nm are being used as delivery systems in pharmaceutical industries. To deliver drugs, the nanoemulsions are made using pharmaceutical surfactants, which are generally regarded as safe (GRAS). Hydrophilic drugs are preferably transported by water-in-oil nanoemulsions, whereas oil-in-water nanoemulsions are used for delivery of lipophilic drugs. Nanoemulsions as the drug carrier have major benefits of improved drug loading capacity, solubility and bioavailability, reduced patient variability, controlled target release, and protection from enzymatic degradation. The main applications of nanoemulsion in drug delivery include cosmetics and transdermal delivery of drug, cancer therapy, vaccine, delivery system, nontoxic disinfectant cleaner, cell culture technology, prophylactic in bioterrorism attack, and improved delivery of poorly soluble drug (oral, ocular, optic, intestinal, parenteral, and pulmonary drug delivery). Moreover, nanoemulsions are utilized in the systems to deliver either recombinant proteins or inactivated organisms to a mucosal surface to produce an immune response. Influenza vaccine and HIV vaccine are some examples (Chime et al. 2014).

On account of the better stability as compared to conventional emulsion and high solubilization of drug molecules, nanoemulsions serve as promising drug delivery vehicles. Nanoemulsions have been widely used as oral drug delivery and transdermal drug delivery system. Various researches are ongoing on the application of nanoemulsions in ocular, pulmonary, nasal, vaginal, and parenteral drug delivery. However, during designing of the drug delivery system, proper selection of oils, surfactants, and cosurfactants need to be considered because the solubility of the drug is affected by these factors (Azeem et al. 2009).

The specific formulation development can also be affected by the solubility of the drug in the oil phase. Edible oil is not frequently used in emulsion formation, because of its inability to readily dissolve large amount of lipophilic drugs. Moreover, the use of low drug soluble oil will lead to the incorporation of high amount of oil to acquire the target drug dose and it will also increase the concentration of surfactants, which might increase the toxicity. The drug's lipophilicity incorporated in lipid nanoemulsions has been considered as a major factor for the biodistribution of the drug. When the lipophilicity is low, the drug is released faster to the blood from the oil droplet, whereas if the lipophilicity is high, the drug will be retained in the oil droplet.

A study conducted by Dierling and Cui (2005) found that Primaquine, a drug used to treat various stages of parasitic malaria, showed an improved uptake in the liver when encapsulated in nanoemulsions as compared to the free solution form. Drugs such as 5-aminolevulinic acid, diclofenac, testosterone derivatives, Propofol, CoQ10, NB-001, cyclosporine, and so on, have been encapsulated in nanoemulsions form for clinical research purposes (Singh et al. 2017).

8.11 Nanoemulsions in Cosmetics Industry

Nanoemulsions have become an interesting method of delivery for cosmetics and optimized dispersion of active ingredients in skin layers. On account of the lipophilic characteristics, nanoemulsions have been intensively used in the transportation of lipophilic drugs. The high surface area of nanoemulsion is a leading factor that has been widely used in cosmetic industries. In addition, some other advantages would be less possibility of creaming, sedimentation, flocculation, and coalescences. Nanoemulsions can effectively transport the active ingredients through the skin and increase their concentration in the skin. An increasing attention has been given toward the production of nanoemulsion without emulsifiers based on polyethylene glycol (PEG). Some of major applications of this type of nanoemulsions are moisturized tissues, wet wipes for baby care, and makeup removal products (Sharma and Sarangdevot 2012).

A nanoemulsion composed of avocado oil (5%), titanium dioxide (0.25%), and octyl methoxycinnamate (1%) was used to produce sunscreen and it showed a slow and sustained release of octyl methoxycinnamate for a period of 4 h (Silva et al. 2013). Recently, industries are using nanoemulsion to develop cosmetic products such as nanocream, nanogel, deodorants, shampoo, and skin care products because of good sensorial properties such as rapid penetration, hydration, and smooth texture.

8.12 Nanostructure Material as Target Delivery System

Nanobiotechnology, which is an interdisciplinary field of chemistry, physics, medicine, and engineering, is changing the aspect of delivery systems of bioactive compounds and nanodevices. The development of nanostructures for the encapsulation and delivery purpose has also enabled the scope of specific or target delivery and controlled release mechanisms. These structures can be tailor made for specific application: drug encapsulation (interaction between carrier and drug), tune release kinetics (cleavage of covalent spacer), regulate biodistribution (molecular weight, addition of targeting group), biodegradability (backbone, spacer), biocompatibility, and minimize toxicity, thereby promoting the therapeutic index of the given drug. Nanostructure-oriented target delivery overcomes the limitation of short life span, poor solubility, stability, and potential immunogenicity associated with macromolecular delivery systems. In addition, nanostructure delivery systems, with the size in nanometer scale, are so small that they can overcome the barrier of cell membranes and deliver the drug to the target cell with lower chance of undesired removal by spleen or liver and lower risk of uptake by reticuloendothelial system. Higher surface area to volume ratio of nanostructures has the advantages of higher dissolution rate, which improves the solubility of the loaded bioactive compounds (Goldberg et al. 2007).

Nanoencapsulation of drugs (nanomedicines) has the capability of enhancing the medication adequacy, specificity, and promoting ingestion into the target tissue, bioavailability, maintenance time, and change of intracellular entrance. Polymeric nanoparticles (poly-D, L-lactide-co-glycolide, polylactic corrosive, poly-ε-caprolactone, poly-alkyl-cyanoacrylates, chitosan, and gelatin) are broadly utilized for the nanoencapsulation of different valuable bioactive particles and restorative medications to prepare nanomedicine. Such nanoparticles help in controlled and target delivery, subcellular size, and biocompatibility with tissues and cells. Aside from this, these nanomedicines are steady in blood, nonlethal, nonthrombogenic, nonimmunogenic, noninflammatory, do not initiate neutrophils, biodegradable, stay away from reticuloendothelial framework, and are relevant to different particles, for example, drugs, proteins, peptides, or nucleic acids. Nanostructured drug delivery systems appear to be a reasonable and promising system for the biopharmaceutical industry compared to the conventional drug delivery system because of the increased bioavailability, dissolvability, and porousness of drugs along with the target delivery and controlled release (Kumari et al. 2010).

Target drug delivery utilizing the nanostructures can be achieved by two mechanisms: (1) active or (2) passive. Conjugation of drug-loaded nanostructure with specific tissue or cell ligand assists in active delivery, whereas in passive delivery the carrier system reaches to the target site indirectly, for instance, drug-loaded nanoparticles passively target tumors cells. Nanoparticles have also been designed to deliver drugs across several biological barriers to overcome limitation of some drugs such as antineoplastic, antiviral drugs, and others that are not able to cross the blood-brain barrier (BBB) (Singh and Lillard 2009). Drug-release mechanism from nanostructure drug delivery system is dependent on drug solubility, desorption of bound or adsorbed drug, drug diffusion through nanoparticle matrix, degradation of nanoparticle matrices, and combination of both erosion and diffusion. The drug-release mechanism is largely controlled by the diffusion process in case where matrix erosion is slower than the drug diffusion rate. The initial rapid drug release may occur when the drug is weakly adsorbed to nanoparticles, whereas sustained release is observed if the drug is loaded by the incorporation methods. Further, addition of polymer in nanoparticles assists in a controlled diffusion drug release. For instance, incorporation of ethylene oxide–propylene oxide block copolymer (PEO–PPO) to chitosan results in reduced drug–matrix interaction and increased drug release. Different methods such as side-by-side diffusion, dialysis bag diffusion, and agitation followed by centrifugation and filtration can be used to monitor the drug release (Singh and Lillard 2009).

References

Abbas, S., K. Hayat, E. Karangwa, M. Bashari, and X. Zhang. 2013. An overview of ultrasound-assisted food-grade nanoemulsions. *Food Engineering Reviews* 5:139–157.

Aboofazeli, R. 2010. Nanometric-scaled emulsions (Nanoemulsions). *Iranian Journal of Pharmaceutical Research* 9:325–326.

Ahmed, K., Y. Li, D. J. McClements, and H. Xia. 2012. Nanoemulsion- and emulsion-based delivery systems for curcumin: *Encapsulation and Release Properties.Food Chemistry* 132:799–807.

Anal, A. K. and W. F. Stevens. 2005. Chitosan-alginate multilayer beads for controlled release of ampicillin. *International Journal of Pharmaceutics* 290:45–54.

Anal, A. K., A. Tobiassen, J. Flanagan, and H. Singh. 2008. Preparation and characterization of nanoparticles formed by chitosan-caseinate interactions. *Colloids and Surfaces B: Biointerfaces* 64:104–110.

Anal, A. K. and H. Singh. 2007. Recent advances in microencapsulation of probiotics for industrial applications and targeted delivery. *Trends in Food Science and Technology* 18:240–251.

Anal, A. K., W. F. Stevens, and C. Remuñán-López. 2006. Ionotropic cross-linked chitosan microspheres for controlled release of ampicillin. *International Journal of Pharmaceutics* 312:166–173.

Anandharamakrishnan, C. 2014. *Techniques for Nanoencapsulation of Food Ingredients.* New York: Springer.

Antoine, D., S. Georgé, C. M. G. C. Renard, and D. Page. 2013. Physicochemical parameters that influence carotenoids bioaccessibility from a tomato juice. *Food Chemistry* 136:435–441.

Azeem, A., M. Rizwan, F. J. Ahmad et al. 2009. Nanoemulsion components and selection: A technical note. *American Association of Pharmaceutical Science Technology* 10:69–76.

Chime, S. A., F. C. Kenechukwu, and A. A. Attama. 2014. Nanoemulsions- advances in formulation, characterization and applications in drug delivery. In *Application of Nanotechnology in Drug Delivery* (Ed.) A. D. Sezer, InTech, DOI: 10.5772/58673.

Chow, S. F., K. Y. Wan, K. K. Cheng et al. 2015. Development of highly stabilized curcumin nanoparticles by flash nanoprecipitation and lyophilization. *European Journal of Pharmaceutics and Biopharmaceutics* 94:436–449.

Dierling, A. M. and Z. Cui. 2005. Targeting primaquine into liver using chylomicron emulsions for potential vivax malaria therapy. *International Journal of Pharmaceutics* 303:143–152.

Ezhilarasi, P. N., P. Karthik, N. Chhanwal, and C. Anandharamakrishnan. 2013. Nanoencapsulation techniques for food bioactive components: A review. *Food and Bioprocess Technology* 6:628–647.

Fan, Y., J. Yi, Y. Zhang, Z. Wen, and L. Zhao. 2017. Physicochemical stability and in vitro bioaccessibility of β-carotene nanoemulsions stabilized with whey protein-dextran conjugates. *Food Hydrocolloids* 63:256–264.

Gharibzahedi, S. M. T. 2017. Ultrasound-mediated nettle oil nanoemulsions stabilized by purified jujube polysaccharide: Process optimization, microbial evaluation and physicochemical storage stability. *Journal of Molecular Liquids* 234:240–248.

Goldberg, M., R. Langer, and X. Jia. 2007. Nanostructured materials for applications in drug delivery and tissue engineering. *Journal of Biomaterial Science. Polymer Edition* 18:241–268.

Gomez-Estaca, J., T. A. Comunian, P. Montero, R. Ferro-Furtado, and C. S. Favaro-Trindade. 2016. Encapsulation of an astaxanthin-containing lipid extract from shrimp waste by complex coacervation using a novel gelatin-cashew gum complex. *Food Hydrocolloids* 61:155–162.

Gupta, A., H. B. Eral, T. A. Hatton, and P. S. Doyle. 2016. Nanoemulsions: Formation, properties and applications. *Soft Matter* 12:2826–2841.

Hu, D., C. Lin, L. Liu, S. Li, and Y. Zhao. 2012. Preparation, characterization, and *in vitro* release investigation of lutein/zein nanoparticles via solution enhanced dispersion by supercritical fluids. *Journal of Food Engineering* 109:545–552.

Khalid, N., G. Shu, B. J. Holland, I. Kobayashi, M. Nakajima, and C. J. Barrow. 2017. Formulation and characterization of O/W nanoemulsions encapsulating high concentration of astaxanthin. Paper in press *Food Research International*.

Kosaraju, S. L. 2005. Colon targeted deliverys sytems: A review of polysaccharides for encpasulation and delivery. *Crtiical Reviews in Food Science and Nutrition* 45:251–258.

Kumari, A., S. K. Yadav, and S. C. Yadav. 2010. Biodegradable polymeric nanoparticles based drug delivery systems. *Colloids and Surfaces B: Biointerfaces* 75:1–18.

Lakkis, J. M. 2016. Introduction. In *Encapsulation and Controlled Release Technologies in Food System* (Ed.) J. M. Lakkis, pp. 1–13. Chichester, England: John Wiley and Sons.

Liu, X. J. Bi, H. Xiao, and D. J. McClements. 2015. Increasing carotenoid bioaccessibility from yellow peppers using excipient emulsions: Impact of lipid type and thermal processing. *Journal of Agriculture and Food Chemistry* 63:8534–8543.

Mao, L. and S. Miao. 2015. Structuring food emulsions to improve nutrient delivery during digestion. *Food Engineering Reviews* 7:439–451.

Mason, T. G., J. N. Wilking, K. Meleson, C. B. Chang, and S. M. Graves. 2006. Nanoemulsions: Formation, structure, and physical properties. *Journal of Physics: Condensed Matter* 18:635–666.

McClements, D. J. 2010. Design of nano-laminated coatings to control bioavailability of lipophilic food components. *Journal of Food Science* 75:30–42.

McClements, D. J. and Y. Li. 2010. Structured emulsion-based delivery systems: Controlling the digestion and release of lipophilic food components. *Advances in Colloid and Interface Science* 159:213–228.

Mishra, M. K. 2016. overview of encapsulation and controlled release. In *Handbook of Encapsulation and Controlled Release* (Ed.) M. Mishra, pp. 4–15. Boca Raton, FL: CRC Press.

Moelants, K. R. N., R. Cardinaels, S. V. Buggenhout, A. M. V. Loey, P. Moldenaers, and M. E. Hendrickx. 2013. A review on the relation between particle properties and rheological characteristics of carrot-derived suspensions. *Comprehensive Reviews in Food Science and Food Safety* 13:241–260.

Nagao, A., E. Kotake-Nara, and M. Hase. 2013. Effects of fats and oils on the bioaccessibility of carotenoids and vitamin E in vegetables. *Bioscience,Biotechnology and Biochemistry* 77:55–60.

Nasrin, T. A. A. and A. K. Anal. 2014. Resistant starch II from culled banana and its functional properties in fish oil emulsion. *Food Hydrocolloids* 35:403–409.

Nasrin, T. A. A. and A. K. Anal. 2015. Enhanced oxidative stability of fish oil by encapsulating in culled banana resistant starch-soy protein isolate based microcapsules in functional bakery products. *Journal of Food Science and Technology* 52:5120–5128.

Oliveira, D. R. B., M. Michelon, G. de Figueiredo Furtado, R. Sinigaglia-Coimbra, and R. L. Cunha. 2016. β-Carotene-loaded nanostructured lipid carriers produced by solvent displacement method. *Food Research International* 90:139–146.

Salvia-Trujillo, L., C. Qian, O. Martín-Belloso, and D. J. McClements. 2013. Influence of particle size on lipid digestion and β-carotene bioaccessibility in emulsions and nanoemulsions. *Food Chemistry* 141:1472–1480.

Salvia-Trujilo, L., O. Martín-Belloso and D. J. McClements. 2016. Excipient nanoemulsion for improving oral bioavailabilty of bioactives. *Nanomaterias* 6:1–16.

Sharma, S. and K. Sarangdevot. 2012. Nanoemulsions for cosmetics. *International Journal of Advanced Research in Pharmaceutical and Bio Sciences* 2:408–415.

Silva, F. F. F., E. Ricci-Júnior, and C. R. E. Mansur. 2013. Nanoemulsions containing octyl methoxycinnamate and solid particles of TiO_2: Preparation, characterization and in vitro evaluation of the solar protection factor. *Drug Development and Industrial Pharmacy* 39:1378–1388.

Singh, R. and J. W. Lillard. 2009. Nanoparticle-based targeted drug delivery. *Experimental and Molecular Pathology* 86:215–223.

Singh, Y., J. G. Meher, K. Raval et al. 2017. Nanoemulsion: Concepts, development and applications in drug delivery. *Journal of Controlled Release* 252:28–49.

Solans, C. and M. J. García-Celma. 2017. Microemulsions and nano-emulsions for cosmetic applications. In *Cosmetic Science and Technology* (Eds.) K, Sakamoto, R. Y. Lochhead, H. I. Maibach, and Y. Yamashita, pp. 507–518. Amsterdam, the Netherlands: Elsevier.

Solè, I. and C. Solans. 2013. Nanoemulsions. In *Encyclopedia of Colloid and Interface Science* (Ed.) T. Tadros, pp. 733–747. Berlin, Germany: Springer.

Teo, A., S. J. Lee, K. K. T. Goh, and F. M. Wolber. 2017. Kinetic stability and cellular uptake of lutein in WPI-stabilised nanoemulsions and emulsions prepared by emulsification and solvent evaporation method. *Food Chemistry* 221:1269–1276.

Ünlüsayin, M., N. G. Hădărugă, G. Rusu, A. T. Gruia, V. Păunescu, and D. I. Hădărugă. 2016. Nano-encapsulation competitiveness of omega-3 fatty acids and correlations of thermal analysis and Karl Fischer water titration for European anchovy (Engraulis encrasicolus L.) oil/β-cyclodextrin complexes. *LWT—Food Science and Technology* 68:135–144.

Yildirim, S. T., M. H. Oztop, and Y. Soyer. 2017. Cinnamon oil nanoemulsions by spontaneous emulsification: Formulation, characterization and antimicrobial activity. *LWT—Food Science and Technology* 84:122–128.

Zou, L., B. Zheng, W. Liu, C. Liu, H. Xiao, and D. J. McClements. 2015. Enhancing nutraceutical bioavailability using excipient emulsions: Influence of lipid droplet size on solubility and bioaccessibility of powdered curcumin. *Journal of Functional Foods* 15:72–83.

9

Nanoparticles, Biointerfaces, Molecular Recognition, and Biospecificity

9.1 Nanoparticles and Biointerface

Nanotechnology has emerged as a new branch of science that has broad applications in many fields such as energy production, industrial production processes, biomedical applications, and so on. A wide range of nanomaterials, especially nanoparticles, are developed with unique and desirable compositions and properties allowing them to be used in novel techniques (Luo et al. 2006; Wang and Wang 2014). Nanoparticles are considered as ultradispersed solid supramolecular structures with at least 1 dimension less than 1 µm, which are generally obtained as nanospheres (matrix types) or nanocapsules (reservoir types) by various methods of preparations (Couvreur 2013). It has been found that nanoparticles exhibit some of the unique properties as compared to bulk materials. The small size and increased surface area of nanoparticles offered unique physical, chemical, electrical, and optical properties that can be utilized for developing the enhanced sensing devices (Luo et al. 2006).

Nanoparticles can either be from natural origin or manufactured synthetically. Some of the naturally based nanoparticles are proteins, polysaccharides, viruses, iron oxyhydroxides, aluminosilicates, metals, and so on, and they are produced by the processes such as weathering, volcanic eruptions, wildfires, or microbial processes (Heiligtag and Niederberger 2013). Nanoparticles can occur in amorphous or crystalline form and they can be carrier for gases or liquid droplets. On account of their distinct features such as larger surface area and quantum size effects, nanoparticles can be considered as a different state of matter from solid, liquid, and gaseous states (Buzea et al. 2007).

A broad range of nanoparticles with different sizes, shapes, compositions, and functionalities are available for different applications. Moreover, these nanoparticles can be further fabricated with various methods such as nanoprecipitation and lithography for developing polymeric nanoparticles. The nanoparticles that are relevant for the application in biological researches

can be listed as liposomes, albumin-bound nanoparticles, polymeric nanoparticles, iron oxide, quantum dot, gold nanoparticles (AuNPs), carbon
nanotubes (CNTs), grapheme-based nanoparticles and magnetic nanoparticles (Wang and Wang 2014; Zeng et al. 2016). Metal-based nanoparticles
are suitable for improving the electrochemical reactions and electrode transfer because of their excellent conductivity and catalytic properties. Oxide
nanoparticles have better biocompatibility, which permit them to be used in
immobilization of biomolecules, whereas semiconductor nanoparticles are
used in electrochemical analysis as labels or tracers (Luo et al. 2006).

The two significant factors of nanoparticles that separate them out from
the bulk materials are surface effects and quantum effects. Surface effects
result in smooth properties' scaling because of the fraction of atoms at the
surface, whereas quantum effects cause discontinuous behavior because of
quantum confinement effects in materials with delocalized electrons. On
account of these factors, nanoparticles have enhanced electrical, mechanical,
optical, and magnetic properties along with the chemical reactivity. The particle with 60 nm diameter is approximately 1000 fold more reactive than
particle with 60 μm because of the larger surface area for chemical reactions. Similarly, nanoparticles require lower binding energy per atom as the
atoms at the surface of small particles will have few neighbors than the bulk
particles. The reduction in binding energy causes a decline in the melting
point of particle, for example, there is more than 300° drop in the melting
temperature when the gold particle size is reduced to 3 nm from the bulk
gold (Buzea et al. 2007).

In addition, there are many advantages of nanoparticles over the bulk ones
for various applications and they can be listed as follows:

1. Products based on nanoparticles exhibit great sensitivity and specificity for production of electrochemical and other types of biosensors
   (Zeng et al. 2016).

2. Nanoparticles can bind, absorb, and transport molecules such as
   DNA, RNA, small molecule drugs, proteins, and probes in an efficient way because of their small size and huge surface-to-volume
   ratio (Thakor and Gambhir 2013).

3. They have better stability, high holding capacity, ability to integrate
   both hydrophilic and hydrophobic compounds, and compatible with
   many administration routes (Thakor and Gambhir 2013).

4. Nanoparticles (especially AuNPs) have low toxicity, high absorption
   coefficient, scattering flux, conductivity, luminescence and ability to
   enhance electromagnetic fields, and enhanced fluorescence (Hutter
   and Maysinger 2013).

5. Nanoparticles are suitable for various administration routes such
   as oral, systemic, pulmonary, and transdermal for drug targeting,
   improvement of bioavailability, bioactivity, and stability of drugs.

6. Nanoparticles have the ability to permeate through cells that helps in the efficient drug delivery to the target site.

7. Polymer nanoparticles can be employed for controlled, sustained, and targeted release of the drug at the site with enhanced therapeutic efficiency and less side effects.

8. Nanoparticle-based antitumor agents have indicated prolonged drug retention in the tumor cells, reduced growth of tumor, and increased survival of species with tumor (Mu and Feng 2003).

These nanomaterials, when come in contact with different biological systems such as organelles, proteins, lipids, DNA, and membranes, form a series of interfaces. These interfaces between the biological systems and nanomaterials are known as nanobiointerfaces. This interface involves the dynamic physicochemical interactions, kinetics, and thermodynamic exchanges between the surfaces of nanomaterial and biological components (Zhang et al. 2015). Surfaces and interfaces play important role in most of the biological reactions (Castner and Ratner 2002). Therefore, the study of biointerface or biosurface is essential to understand the principles and working mechanisms of biomaterials for their utilization, both structurally and functionally. The adhesion and spatial configurations of biomolecules are mostly affected by the specific chemical functional groups and micropatterns of biointerface/biosurface. The biointerface study of graphene oxide with lipid membranes, nucleic acids, proteins, and so on could be beneficial in the development of biosensors for the field of medicine, food, and beverage industries. The good connection between the biomaterials can be set up by a stable and firm adhesion on the interface. The biointerface/biosurface interactions are peptide interactions, lipid bilayers, protein antitumor, interactions between proteins, and hydroxylapatite interaction with biomolecules (Tang and Zhang 2016).

The organic and synthetic worlds meet at the interface between nanomaterials and biological systems giving the issue related with the safe application of nanotechnology and nanomaterial design for the applications in biological systems. The three dynamic interacting components that exist between nanobio interfaces are as follows (Figure 9.1):

1. The nanoparticle surface which is characterized by its physicochemical compositions.

2. The solid–liquid interface and the changes that occur when the particle interacts with the components of the surrounding medium

3. The solid–liquid interface contact zone with biological substrates

The surface properties of nanoparticles are influenced by the shape and angle of curvature, chemical composition of the material, surface functionalization, porosity and surface crystallinity, roughness, heterogeneity, and

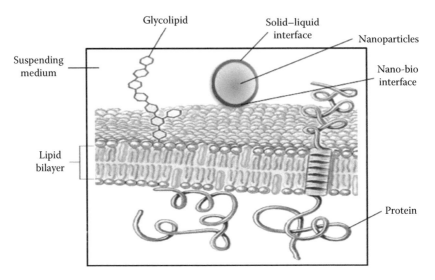

FIGURE 9.1
Interface between nanoparticle and biological systems. (From Nel, A.E. et al., *Nat. Mat.*, 8, 543–557, 2009; Nelson, D.L. and Cox, M.M., *Lehninger Principles of Biochemistry*, W. H. Freeman, New York, 2012. With Permission.)

hydrophobicity/hydrophilicity. The different characteristics of the suspending media such as pH, temperature, ionic strength, and the presence of large organic molecules or detergents determine the surface properties that include stability/biodegradability, zeta potential, aggregation of particle, dispersion and dissolution characteristics, hydration, and valence of the surface layer (Nel et al. 2009).

At the nanobiointerface, there is interaction of nanomaterials with ions, metabolites, various proteins, lipids, DNA, microbial products, and so on in the suspending medium. This may result in the alteration of the structure and function of the biological components in the suspending medium, and affect cellular functions. The nanomaterials might interact with the biochemical constituents of plasma membrane after getting adsorbed on the cell surface (Zhang et al. 2015). Leroueil et al. (2008) found that the cationic nanoparticles with different properties can cause lipid bilayer disruption on biological membranes. Moreover, if the nanomaterials are taken up further, they may have cytotoxic effect inside the cell. The nanobiointeractions can also transform the nanomaterial properties such as dispersion ability, dissolution, surface charge, and surface chemistry. The electrostatic properties of surface and toxicity of nanomaterials can be affected by the ions present in the suspending medium (Zhang et al. 2015).

9.2 Application of Nanobiointerface in Knowing the Interaction between Surface and Biomolecules

9.2.1 Nanobiointerface for Targeting and Therapeutic Delivery

The endogenous transport mechanisms at the cellular level operate on a scale of nanometers. It gives an option for engineered nanomaterials or particles of cellular or subcellular levels in cells, tissues, or organs of living organisms, which can be precisely delivered to targeted locations for diagnostic or therapeutic purpose. Therefore, this technique has evolved as a major science of nanoengineering of biomolecules to deliver drugs, genes, and therapies to treat diseases. The barrier-crossing characteristics of nanoparticles are mainly determined by their polarity and hydrophobicity. Depending on the exposure route, the surface of the nanoparticles initially comes in contact with the biological fluids such as blood, bile, lung fluids, and so on. For example, injection of nanoparticles intravenously will expose them to the blood plasma that contains 3700 different proteins and many other complex biomolecules. These biomolecules competitively try to bind with the surface of the nanoparticles.

The selectivity and specificity of the interactions between nanoparticles and biomaterials are greatly influenced by the size and shape of nanoparticles. Similarly, the nanobiointerface is affected by the surface characteristics and state of the nanoparticles. This may be because of two main reasons:

   a. Even though there are strong interactions between particle and the surrounding, only the nanoparticle of correct size range will get engaged with the biomaterials.
   b. A large interfacial area is available for the interaction as compared to the same mass of bulk material.

The above-mentioned two principles determine the complexities and implications in targeted delivery of drugs to cells (Mahon et al. 2012).

9.2.2 Targeting Strategies for Drug Delivery at Cell Level and Organ or Tissue Level

The approach of targeting is of utmost importance in the delivery of therapeutic effects to prevent the accumulation of drugs on nonspecified sites in the cells or organs. The specific target success will achieve a complete intended therapeutic effect. This is a realization of all the advantages of nanoparticles. The typical nanoparticles employed in the treatment of cancer cells are liposomal. However, there is a growing interest in using the mixture of albumin and taxol because of their ability to inhibit the opsonization of nanoparticles by forming the macrophages. In the modern approach of cancer treatment, the proteins or receptors that are overexpressed in the cancer cells are specifically targeted (Mahon et al. 2012). For example, folic acid conjugated albumin nanospheres (FA–AN) have been recently developed to provide actively targetable drug delivery system for improved drug targeting of cancer cells with the minimal side effects. The nanosphere is prepared by conjugating folic acid onto the surface of albumin nanosphere using some catalysts (Shen et al. 2011).

At the cellular level, the nanoparticles with the biomolecules should reach to the cells to be targeted where there is overexpression of the receptors of those particles. Human transferrin is a glycoprotein (79 kDa), which binds with the transferrin receptor resulting in endocytosis. It has been extensively used as a cancer targeting agent (Thorstensen and Romslo 1990; Wang et al. 2010). Most of the cancer cell types go through cell division at a higher speed that will demand additional iron (for synthesis of heme) leading to overexpression of the transferrin receptor (Daniels et al. 2006).

The in-depth understanding of interaction in nanobioconjugates and nanobiointerface is well explained by the *in vitro* study of nanoparticle–transferrin conjugates. At the cellular level, the nanoparticles made of dextran stabilized with iron oxide, which is functionalized with transferrin showed different trafficking behavior and particle internalization rate when they are covalently bonded with different linkers. These transferring-conjugated nanoparticles were trafficked to unconjugated proteins in different ways and the protein was recycled back.

At the tissue or organ-level targeting, there are two tumor-targeting strategies: (1) passive and (2) active, of engineered nanoparticles of which both the methods ensure that the nanoparticles reach the tumor vicinity. Both the targeted and nontargeted nanoparticles reach the tumor site by passive targeting means or by enhanced permeation and retention effect (EPR). Following this, the internalization of the tumor cell is initiated by surface ligands for active targeting, which should have specific conformation, higher rate of cellular internalization, and higher affinity to the receptors. Then, the cell internalization occurs and the effect of drugs introduced will occur (killing of tumor cells). Sometimes, because of the disruption of targeting ligand or nonspecific biomolecular adsorption from the environment, the success rate of targeting may be lower. Therefore, these two strategies should be combined along with the development of nanoparticles that can functionalize for targeting both tumor cells and tumor vessels for the overall improvement of the targeting efficiency (Mahon et al. 2012).

For the maximum delivery of nanoparticles in the target site, the nanoparticles should have the ability to skip detection by immune system and avoid opsonization, which causes the removal of particles from circulation. Surface modification can be done to avoid the opsonization by coating nanoparticles with polyethylene glycol (PEG). On account of the combination of steric hindrance and hydrophilicity, this will reduce nonspecific protein binding and prevent adsorption of serum and IgG (Zheng et al. 2004; Li and Huang 2010).

9.2.3 The Protein (Biomolecule) Corona

After the exposure to biological environment, the biomolecular corona is formed on the surface of nanoparticles. It involves the exchange of proteins between nanoparticle surface and plasma, between nanoparticle surface and cell surface, and between nanoparticles and the high-affinity free protein molecules in the medium, which can compete for the cell surface (Walczyk et al. 2010).

9.2.4 Electrochemical Nanobiosensors for Food Safety

The food quality and safety are being challenged by pesticide residues, veterinary drug residues, illegal food additives, organic compounds, heavy metals, food pathogens, and toxins. This may lead to food poisoning and food-borne illness that will have negative impact on the health of humans and also on the food business. Therefore, there was an increasing demand of technology that can strictly analyze and control these compounds in food. This led to enormous research on the food sensors and finally development of biosensors to ensure food safety.

Nanomaterial-based electrochemical biosensors possess great potential to be used in the detection of contaminant in the complex food system because of the unique chemical and electrical properties of nanomaterials. Along with the rapid-response reaction, these biosensors have better sensitivity and selectivity. Many nanomaterials possess synergistic effect on the catalytic activity and transduction of signal toward target molecule/compounds, which improve the specific selectivity between molecules. The major three categories of analytes in food safety are (1) food additives, (2) chemical contaminants, and (3) microbial contaminants. The overconsumption of these additives and contaminants may cause different health disorders such as insomnia, anxiety, and irritability, and may degrade the overall food quality.

The nanomaterials based on carbon, grapheme, metal and metal oxide, magnetic, and molecularly imprinted polymers are mostly used in the electrochemical biosensors for the application in the food safety, which is shown in Table 9.1 (Zeng et al. 2016).

TABLE 9.1

Application of Electrochemical Nanobio Sensors in Food Safety

Nanomaterials	Analytes	Sample
Carbon Nanotubes		
MIP–AuNPs–CNTs	Glyphosate and glufosinate	Soil, human serum
MWNTs@MIP	Amoxicillin	Milk, honey
NGR–NCNTs	Caffeine and vanillin	Food
ZnO/Pt/MWNTs	Nitrite	Pickled food
AuNPs/rGO/CNTs	Bisphenol A	Food
Graphene		
AuNPs/rGO	Sunset yellow AuNPs/rGO	Soft drinks
GQDs	Pb^{2+}	Water
Au–PEDOT/rGO	Caffeic acid	Red wine
GR/AuNPs	Aflatoxin B1	Spiked food
Metal		
AuNPs/CdTe/Chitosan	Monocrotophos	Food
AChE	Malathion	Food
Magnetic		
CNTs@f–Fe_3O_4	Chlorpyrifos	Aquatic sample
Fe_3O_4/MWNTs	Malathion	Food
Fe_3O_4–GO–SO_3H	Doxorubicin hydrochloride	Plasma
Fe_3O_4@rGO	Amaranth	Fruit drinks
Fe_3O_4–GO	Hydroquinone	Tap water

Source: Adapted from Zeng, Y. et al., *J. Electroanal. Chem.*, 781, 147–154, 2016.
MIP—molecularly imprinted polymer; CNTs—carbon nanotubes; MWNTs—multiwalled carbon nanotubes; Nanoparticles—NPs; GO—graphene oxide; NGR–NCNTs—nitrogen-doped graphene and nitrogen-doped carbon nanotubes; rGO—reduced graphene oxide; GR—graphene; GQDs—graphene quantum dots; PEDOT—poly(3,4-ethylenedioxythiophene); f-Fe_3O_4–flake-like Fe_3O_4.

9.3 Molecular Recognition

The understanding of molecular recognition is vital in different application fields such as therapeutics, chemical catalysis, and sensors (Zhang et al. 2013). Molecular recognition involves selection and binding of the substrates by the receptor molecule; however, mere binding is not considered as the recognition. It can be defined as a binding phenomenon with a specific purpose; for example, receptors act as ligand with a purpose (Lehn 1973). In another word, molecular recognition in biological system is the process of interaction of macromolecules with each other or other small molecules through noncovalent bond that results in the formation of specific complex. This process comprises two significant features: (1) specificity (differentiates the highly specific binding molecule from the less specific ones) and (2) affinity (determines that even a low concentration of high-affinity specific interacting molecules have more effect than a high concentration of weakly interacting molecules). In fact, molecular recognition is a part of complex and functionally important mechanism that encompass many crucial elements of life such as different metabolism, self-replication, and processing of information. The molecular recognition can be well illustrated in the DNA replication process that occurs before cell division. Replication of DNA is carried out by a series of enzyme-catalyzed reactions that depend on the specific molecular recognition and binding between the specific enzyme and DNA strand/segments. In addition, it also helps in the regulation of metabolic network of numerous chemical reactions that occur parallelly at the same time. A sequence of recognition, binding, and breakdown process occurs in the cellular system. A small molecular messenger is initially recognized, then transmitted through the membrane receptor segment and finally accomplished by the functional response of a binding cell (Du et al. 2016).

9.3.1 Principle of Molecular Recognition

Molecular recognition demonstrates complementary lock and key type fit in between molecules. *Lock* signifies the host structure that can be any gap or crack on the surface of a macromolecule or a hollow site inside a molecular aggregate, whereas the *key* signifies the guest, which can be a small molecule or a portion of a large molecule, fully or partly embedded by the above empty space. Both lock and key are assumed to be rigid, but in many cases, they are flexible and adopt a new final shape after binding. While binding, the conformational changes in the lock (host) are termed as induced fit and that in the key (guest) is called conformational selection (Harmata and Náray-Szabó 2009).

Hence, molecular recognition takes place when three-dimensional (3D) structure of host and guest molecule fit in terms of steric, electrostatic, and hydrophobic interactions that allow the noncovalent binding. This self-organization process is flexible, highly dynamic, conformational, and induced fit that reduces internal surface areas not being saturated intermolecularly and improves the binding ability. The specificity of a given system is the consequences of multiple noncovalent bonds that are formed because of narrow spatial constraints and higher binding strength and affinity over other molecules that are more or less dissimilar. This is the key phenomena of molecular recognition that permits the receptor molecule to recognize a specific substrate molecule because of high specificity from the system of numerous other molecules (Lammerhofer 2009).

Specific recognition on a molecular level needs a complementarity in the shape or surface of both receptor and substrate that will provide the basis for all molecular recognition phenomena. The nonspecific interactions involving van der Waals or hydrophobic forces can become more specific and stronger because of electrostatic interactions or hydrogen binding. Thermodynamically, when the free energy of receptor–substrate complex is lower than the sum of free energy of unbound receptor and substrate, the molecular recognition takes place. Thus, the reversible association of two or more molecules results in the formation of receptor–substrate complex. In the noncovalent binding, the enthalpies are relatively weak, so the interaction of both enthalpy and entropy is more essential than in the covalent bond formation.

The contribution of TΔS because of the loss of translational and conformational freedom to the free energy becomes significant. For example, the freezing of a freely rotating bond with three equally populated conformations into one conformation will increase ΔG by approximately—TΔS 2.9 kJ/mol. In cyclic structures, less reorganization is required for the binding as compared to acyclic structures. Hence, most of the receptors either artificial or natural belong to cyclic structures. Some of the important factors that are involved in the molecular recognition are as follows:

- Complementing shape of molecules
- Appropriate binding sites for noncovalent interactions (ΔH)
- Rigidity in the structure for the reduction of loss in molecular motion during the formation of supramolecular structure (ΔS) (Konig 1995)

9.3.2 Application of Molecular Recognition

Biomolecular interaction is a major phenomenon in many biological and physiological processes. Specific molecular recognition processes are vital on cell surfaces for the proper functioning of different cellular activities such as cell adhesion, tissue assembly, development of embryo, transduction of signal, and immune response. These recognition processes involve complex interactions between different molecules such as antigens and antibodies, membrane receptor proteins and ligands, carbohydrates and lectins, and cell-adhesion molecules (CAMs) and the extracellular matrix (ECM). When the membrane receptor proteins interact with the specific ligands, they serve as mediators for transmitting the biological signals between the intracellular and extracellular environment of the cell. Similarly, the cell–cell interactions on the cell surfaces are mediated by recognition of complementary carbohydrates by lectins on adjacent cells. These recognition and interactions occurring on the cell surfaces can help in characterization of cellular processes such as cell growth, cell differentiation, formation of junction, and other pathological processes (cellular adhesion, infection, and cancer cell metastasis). Moreover, these interactions aid in assessing the specificity and selectivity of biological entities that can be beneficial in the development of cell-specific bioanalytical devices (Wang and Yadavalli 2014).

For the sensing application, molecular recognition and signal transduction are the key factors. In the context of biological applications, traditionally the naturally existing macromolecules having better specificity and sensitivity such as antibodies are used for the development of sensors. Antibodies are transmembrane protein, which can act as the antigen receptor for the cell and are most suitable for the biosensing because of their wide range of specificities. During the transcription process, the antibodies produced by each B-cells are able to recognize a different epitope and this specificity is further enhanced by mutations. However, their application in sensor development is limited because they are fragile, costly, prone to batch variation, and less biological activity on external treatment. Therefore, there was a need for the development of a method for synthesis of artificial antibodies or similar molecules from polymeric materials in biosensing and this resulted in the discovery of molecularly imprinted polymers and DNA aptamers (Donahue and Albitar 2010; Zhang et al. 2013).

Nanomaterials such as single-walled carbon nanotubes (SWNTs) and rolled cylinders of grapheme have great potential to be used for molecular recognition and signal transduction because of their sensitivity and near-infrared band gap fluorescence without photobleaching threshold. Numerous sensors based on SWNT fluorescence have been developed for detecting β-D-glucose, assorted genotoxins, hybridization of DNA, nitric oxide, divalent metal cations, avidin, and pH (Zhang et al. 2013).

9.3.3 Potential Applications of Biopolymeric Nanoparticles

The unique properties of nanoparticles along with their biocompatibility, low toxicity, and ease of surface modification encourage them to be applied in a wide range of fields from biomedical applications to biochemical, industrial, environmental, and food applications. Some of the well-known applications of nanomaterials include microelectronics, catalytic agent, coatings and adhesives, UV absorbers for sunscreens, synthetic bone, cosmetics, fabrics and their treatments, dental materials, surface disinfectants, diesel and fuel additives, dangerous chemical neutralizers, drug delivery systems, and pharmaceutics. Nanoparticles are expected to improve the diagnosis of disease and their treatment in medical applications, whereas in environmental application, it is used to eliminate the pollutants from water and soil resources (Buzea et al. 2007; Mou et al. 2015; Hannon et al. 2015).

9.3.3.1 Application in Medicine

Successful application of nanotechnology in medical sectors is one of the remarkable achievements for the benefits of humans, animals, and plants to combat diseases and disorders. Most of the cell components are in the dimension of nano size; therefore, there is a wide room and scope for nanotechnology to expand its applications in the medical sector. From the times not so far, nanotechnology has been fruitfully used in the detection and intervention of diseases, serving as drug delivery means and medical diagnostics tools. Studies have shown that these particles have tremendous pharmacological potential in improving synthesis of drugs, carriers, and optimizing the material and reducing the toxicity. On the basis of the survey carried out by European Science and Technology Observatory, 2006, 24 different nanotechnology-based therapeutic drugs have been granted for medical use, wherein liposomal drugs and polymer-drug conjugates being the most prevalent. Today the world is practicing more than 250 nanomedicines in different steps of medical development, particularly in drug delivery applications (Fonseca et al. 2014).

AuNPs have been used to develop the biochips that can detect osteoporosis, analyze the condition of bone, and provide information about the degree of bone damage. Biosensors made up of CNTs can be employed for the diagnosis of bone diseases such as Paget's disease, renal osteodystrophy, and osteoporosis (Garimella and Eltorai 2017). Use of nanocarriers based on liposomes, polymeric nanoparticles (albumin and chitosan), squalene conjugation (squalenoylation), polymeric micelles (PMs), and inorganic nanoparticles (e.g., AuNPs, magnetic iron oxides, CNTs and mesoporous silica) for gemcitabine, and drug for the pancreatic cancer treatment, have shown excellent results in both *in vitro* and *in vivo*, by enhancing the pharmacokinetics, antitumor effect, and the therapeutic effectiveness (Birhanu et al. 2017).

For the screening and treatment of cancer, conventional imaging techniques such as radiographs, ultrasound, computed tomography (CT), and magnetic resonance imaging (MRI) have been used widely. But they are limited to detection of cancer after it develops into a visible physical unit, that is, around 1 cm^3 and at this point there will be around 1 billion cancer cells in the tumor mass. So, the molecular imaging can be useful in detecting the cancer earlier at the molecular level and way before the physical visibility occurs. Nanoparticles have a great potential to be used as molecular imaging agents as they can enhance signal intensity, thus permitting imaging of less numbers of cells at tissue depths and stable imaging signals. For example, superparamagnetic iron oxide for MRI, heavy metal nanoparticles (gold, tantalum, and lanthanide) for CT, single-walled CNTs for photoacoustic, and so on, have been used for imaging preclinical and clinical stage of cancer development (Thakor and Gambhir 2013).

FA–AN can be used for actively targetable drug delivery system for better targeting of drugs to cancerous cells with minimal side effects. Doxorubicin, an antibiotic drug for cancer chemotherapy, was loaded in FA–AN nanoparticles and it was then acted on the tumor cells (HeLa cells). The entrapment of drug in FA–AN nanoparticles resulted in selective killing of the tumor cells as compared to the free drug without entrapment and sustained release of drug that minimized the harmful side effect of doxorubicin (Shen et al. 2011).

9.3.3.2 Application in Food Science

With the invention of pasteurization technique to destroy the food spoiling bacteria (1000 nm) by Louis Pasteur, nanotechnology is creating a revolution in food processing and food quality improvement. Today, nanotechnology has become an indispensable part in the field of food science and it imparts several benefits to both food industries and consumer; however, the potential health risks from those nanomaterials should be considered. The main aim of using nanosized particles in food and related products are to extend the shelf life of food product, improve food production processes, enhance stability, texture and consistency of products uptake, and enhance the nutrient uptake and its bioavailability. Many researchers and food industries are currently working on the applications of nanoscience to reduce or substitute the use of pesticides and antibiotics in food production, to enhance the nutritional value, and to modify the mechanical and sensory properties of food as desired by the consumers. According to Internet databases, around 140 food-related products were identified to contain nanomaterials ranging from organic to inorganic metal and metal oxides (Bouwmeester et al. 2014; Chellarama et al. 2014).

The application of nanoparticles has covered almost all the stages of food production chain from farm to fork. They have been used at the farm, at the manufacturing plant during the processing and packaging of product, and at the storehouse during their storage before the final consumption by consumers. Furthermore, they are also studied for their role in the controlled release of nutrients after digestion, encapsulation efficiency of bioactive compounds, development of biosensors for food pathogens, and smart packaging systems (Joye and McClements 2013).

Many food supplements such as proteins, vitamins, iron, calcium, and other bioactive compounds are available in the nanoform, mainly as nano-encapsulates for their use as food or health supplements. Nanosilica and nanosilver have been used in food applications. Nanosilver has antibacterial properties. Nanosilica, a registered food additive, can be used as a carrier for delivering minerals during food processing (Bouwmeester et al. 2014). Nanodelivery of food bioactives such as antioxidant compounds, probiotics, proteins, polyunsaturated fatty acids, vitamins, and so on, helps to improve the solubility, stability, functionality, and bioavailability, and also allows the controlled release of these constituents (Borel and Sabliov 2014). Nanoparticle possesses numerous advantages for delivery of bioactive compounds in food systems as listed below (Jahanshahi and Babaei 2008):

1. High loading capacity
2. No or less chemical reactions during preparation of nanoparticles
3. Resistant to various processing conditions
4. Easy to manipulate the composition, size, shape, and surface characteristics for effective targeted delivery in the GIT

TABLE 9.2

Nanoencapsulation of Bioactive Compounds for Food Application

Bioactive Compounds	Wall Material	Objective
β-carotene	Ethyl cellulose and methyl cellulose	To enhance stability and bioavailability
Linoleic acid	α- and β-cyclodextrin	To improve thermal stability
Bovine serum albumin	Chitosan, poly (ethylene glycol-ran-propylene glycol)	For control release
Curcumin	Poly (lactide-co-glycolide), PEG-5000	To improve solubility
Astaxanthin	Poly (D, L-lactic acid) and ply (D, L-lactic-coglycolic acid) gelation or tween 20	To improve solubility and bioavailability
Vitamin E	PEG, tween 80	To improve stability, retention capacity, and shelf life
Coenzyme Q10	Poly (methyl methacrylate) and polyvinyl alcohol	To enhance stability, reproducibility, and drug-loading yield

Source: Adapted from Ezhilarasi, P.N. et al., *Food Bioproc. Technol.,* 6, 628–647, 2013.

Few of the food-bioactive compounds that have been transformed to nanoparticles for their application in food are shown in Table 9.2.

On account of their selectivity, sensitivity, specificity, and adaptability, biosensors can be used for wide application in food industry mostly in food safety and quality control. Biosensors for food application have been developed using different types of nanoparticles such as metal, semiconductor, oxide, magnetic, and composite nanoparticles. Biosensors designed with Bismuth nanofilms and peptide nanotubes have been successful in detecting *Escherichia coli* in food samples. Chitosan/titanium oxide nanoparticle-based sensor was developed to analyze ochratoxin (Farah et al. 2016). Sun et al. (2008) designed an electrochemical enzyme immunosensor, which is based on the bioelectrocatalytic reaction of imidazolium cation room-temperature ionic liquid and AuNPs, fabricated on glassy carbon electrode for the detection of aflatoxin B1. A novel biosensor of chitosan nanofiber/AuNPs composite was capable of sensing cholesterol level with high sensitivity and less response time (around 5s) (Gomathi et al. 2011).

Nanomaterial incorporated sensors have been invented to regulate and monitor the internal environment of food products, including temperature, oxygen content, pathogens, and other components. These sensors when applied on the food packaging system can also predict the shelf life of food products. Engineered AuNPs incorporated with enzymes are capable of detecting microbes; zinc and titanium oxide nanocomposites can detect the volatile organic compounds and perylene-based fluorophore nanofibrils can examine the spoilage of meat and fish by measuring gaseous amines. Similarly, nanobarcodes are applied for tagging the food products for assisting in traceability (Chellarama et al. 2014).

Generally, nanoparticles are incorporated in the food packaging system for improving physical and functional properties that will transfer the packaging system into improved, active, or intelligent packaging. Different metal and nonmetal nanoparticles have been exploited in the food packaging system; for example, copper, copper oxide, gold, iron, iron oxide, silicon dioxide, silver, titanium dioxide, titanium nitride, zinc oxide, and so on, as metal-based nanoparticles and clays, protein, polymers, chitosan, poly-lactic acid, and so on, as nonmetal nanoparticles. Silver nanoparticles have been incorporated into storage box, fresh food containers, and salad bowl to impart the antimicrobial properties so as to prevent the food spoilage and extend the shelf life. Nanoparticles of zinc oxide have been entrapped in the plastic wrap to improve the barrier properties and also to integrate the antimicrobial properties. Similarly, nanoclay has been used in beer bottles and films for strengthening the barrier properties.

On account of the possible inherent toxicity of these metal and nonmetal nanoparticles, there was a search for the food-grade nanoparticles. Some of the food-derived compounds such as lecithin, sodium caseinate, curcumin, and vitamin C have been reduced to nanoscale and incorporated into the packaging system to provide antimicrobial functions. The nanomaterial of paprika oleoresin has been found to improve the marinating performance and sensory properties of poultry meat. In addition, corn starch in nanoform has been used to develop ecoplastic that is easily biodegradable (Hannon et al. 2015).

9.3.3.3 Application in Biotechnology

The availability of wide variety of nanomaterials with unique properties in the nanometer size range has ignited great interest in their application in biotechnological systems. However, they strictly require having certain physical, chemical, and pharmacological properties such as chemical composition, uniformity, crystal and surface structure, adsorption properties, solubility, and low toxicity (Tartaj et al. 2005). Various kinds of biosensors that have been developed for the diagnosis, prognosis, and treatment of different kinds of diseases can be summarized as follows:

1. A galactosylated gold–nanoisland biointerface-based biosensor can be used for the identification of hepatic cancer cells in bloods with better specificity and biocompatibility (Liu et.al. 2016).
2. Biosensors, based on the platinum nanoparticles have been used for electrochemical amplification with a single-label response for the detection of low concentration of DNA (Kwon and Bard 2012).

3. Semiconductor nanoparticles such as quantum dots and iron oxide nanocrystals can be linked with tumor-targeting ligands (peptides, antibodies, or small molecules) for targeting the tumor antigens with greater specificity and affinity (Nie et al. 2007).

4. Concentration of xanthine has been quantified in the coffee sample by using biosensors based on carboxylated multiwalled carbon nanotubes (c-MWCNTs)-7,7',8,8' tetracyanoquinodimethane (TCNQ)–chitosan (CHIT) composite (Dalkıran et al. 2017).

5. α–1 antitrypsin (AAT), a biomarker for Alzheimer's disease, can be electrochemically detected by a sandwich-type biosensor. This biosensor was prepared using 3, 4, 9, 10-perylene tetracarboxylic acid/ carbon nanotubes (PTCA–CNTs) as a sensing platform, whereas alkaline phosphatase-labeled AAT antibody functionalized silver nanoparticles (ALP–AAT Ab–Ag NPs) as a signal enhancer (Zhu and Lee 2017).

6. Quantum dots, a highly luminescent semiconductor, based on zinc sulfide-capped cadmium selenide and coupled with specific biomolecules can be utilized for ultrasensitive detection of biological compounds. These nanoscaled bioconjugates are soluble in water and biocompatible. Quantum dots labeled with the protein transferrin resulted in receptor-mediated endocytosis in cultured HeLa cells, and that labeled with immunomolecules identified specific antibodies or antigens (Chan and Nie 1998).

7. A highly sensitive probe was developed from mercaptoalkyl oligonucleotide-modified AuNPs that can detect polynucleotide on the basis of colorimetric principle. When the polynucleotides interact with nanoparticles, polymeric network is formed causing change in the color of the solution (Elghanian et al. 1997).

8. Gold nanoshell can either absorb or scatter incident light with the variation in their size. The photothermal effect produced by optically absorbing gold nanoshells–polymer composite can be used for the controlled release of the drug from the nanoshell–polymer composite (West and Halas 2000).

9.4 Key Parameters for Development of Biopolymeric Nanoparticles

The versatility and unique properties of nanoparticles allow them to be explored for many applications in medicine, food, and biotechnology with number of choices in the composition, loading, and target moieties.

The characteristics such as size, size distribution, surface morphology, surface chemistry, charge, adhesion, stabilization, loading capacity, release behavior, and thermodynamic properties of the nanoparticles affect the *in vitro* and *in vivo* properties of nanoparticles. Size of the nanoparticles and their distribution can significantly affect the effectiveness and safety. Similarly, the surface characteristic can play an important role in the behavior of nanoparticles, their stability, and their interaction with biomolecules (Desai 2012).

Besides these characteristics, there are few other parameters that need to be considered while developing nanoparticles for their nanobio applications.

9.4.1 Biocompatibility of Nanoparticles

Biocompatibility of nanoparticles can be defined as the ability of a nanomaterial to perform its desired function without causing any undesirable effects in the host or its performance (Li et al. 2012). Generally, a high degree of biocompatibility can be attained when particular nanomaterial interacts with the body cell devoid of triggering unacceptable toxicity, immune, and carcinogenic responses. The biocompatibility of materials can be evaluated by considering the following three factors:

- First, biocompatibility is greatly dependent on the anatomical properties of the target unit. So, the same nanoparticle may have different effect or reaction from one site to another. They can be have mild or no effect on one tissue, whereas strong effect on another one depending on their application and target.
- Second, the exposure half-life of nanoparticle is more important than the intrinsic properties of the biomaterials for determining whether it is biocompatible or not.
- Finally, biocompatibility depends on the relativity of the risk and benefit (Naahidi et al. 2013).

Nanoparticles have been implanted in the human body to diagnose and treat the disease and deformations. They have been widely used as drug delivery systems to a target cell. However, biocompatibility of nanoparticles must be understood before they are introduced in the living cells. Studies should be done to understand their impact on the viable cells, including their interactions with natural functionality of healthy cells, toxicity level, absorption by the body cells, and so on. Different studies have shown that the toxicity and biocompatibility of nanoparticles in most cases depend on the composition, size, shape, surface coating, surface properties, dosage, functionalization, and the intended use. Any diagnosis of toxic effect can consequently deteriorate the efficiency of therapy (Markides et al. 2012).

Nanoparticles by virtue of their small size exhibit higher permeability through the cell membrane, nerve cells synapses, blood vessel, and lymphatic system. On account of this permeability, nanoparticles are more effective as drug carrier; however, at the same time it also poses potential threats on human. Besides nanoparticles, nanotubes made up of carbon, silicon dioxide, boron nitrate, and titanium dioxide have unlimited applications in the field of biomedicine (Li et al. 2012). The biocompatibility of nanoparticles can be studied under hemocompatibility, histocompatibility, and immunocompatibility.

a. *Hemocompatibility*: Mostly, nanoparticles are used as vector for delivery of drugs or gene or as biosensors and there is direct contact with blood and particles. Thus, the compatibility with blood should be studied. The hemocompatibility of nanoparticles can be understood by investigating the blood cell aggregation, hemolysis, and the coagulation behavior of blood cells in *in vitro* condition. The blood compatibility of nanoparticles is highly dependent on size, surface area, surface charge, and hydrophobicity or hydrophilicity.

b. *Histocompatibility*: One of the major application of nanoparticles in biomedical is for targeted drug delivery. Hence, rigorous studies on histocompatibility of several representative types of nanomaterials have been done so far. The major nanomaterials include super paramagnetic iron oxides (SPION), dendrimers, mesoporous silica particles, AuNPs, and CNTs. SPION have been considered as one of the biocompatible nanoparticles indicating no severity in *in vitro* and *in vivo* toxic effects. The study in primary human macrophages demonstrated no immunomodulatory effects when cells were exposed to 30 nm dextran-coated SPION. The toxic effect of the dendrimers was found to be generation dependent as higher generation dendrimers were more toxic. Similarly, cationic dendrimers are found to be more toxic than the anionic dendrimers. On account of their higher biocompatibility, silica nanoparticles are being used as drug delivery means and biosensors. They were capable of entering the cell causing no damage to surviving cells (Li et al. 2012).

c. *Immunocompatibility*: Immunocompatibility involves the study of immune response to a nano-based material. In the study of biocompatibility of nanoparticles, it is as important as hemocompatibility and histocompatibility. After entering into the body, nanoparticles can either stimulate or suppress the immune system. This immune response may have positive or negative impact on the functioning of particle for the intended application. The factors that are responsible for immunostimulation or immunosuppression include molecular structure, architecture of folding motifs, degradation products, formulation, package purity, and stability of pharmaceuticals along with dosage, route, and time of administration, mechanism of action, and activity site. Immunosuppression may increase susceptibility to infections caused by bacteria, viruses, fungi, and yeast, and development of skin cancer. However, it can be beneficial in the treatment of autoimmune diseases and in accepting foreign tissues during organ transplants (Naahidi et al. 2013).

9.4.2 Biodegradability and Safety of Nanoparticles

Day-by-day, the application of engineered nanoparticles in different fields (medical sectors, pharmaceuticals, agricultural, food, information sectors, and industrial sectors) of human civilization is growing in a sky rocketing manner. However, the question arises about what will be the consequences of these nanoparticles in the environment and their subsequent effects in the ecosystem in the future. Is it sure that these particles will maintain their normal size, original structures, and reactivity in the environment as in their proposed application? Similarly, is their effect on environmental system different from that of larger particles of the same material? Therefore, answers for these questions should be primarily dug out before introducing any forms of nanoparticle-based devices and technologies in the market (Ray et al. 2009).

The biodegradability of nanoparticle is desired after its intended function when it is introduced into the body. Biodegradable nanoparticles are more preferred than nonbiodegradable ones as they are digested internally and afterward cleared from the body. The materials that can be used for preparing biodegradable nanoparticles are proteins, polysaccharides, polymers such as poly-DL lactide-co-glycolide, polylactic acid, poly-ε-caprolactone, chitosan, gelatin, and so on. Besides, the nature of raw material, experimental condition, target site, and animal species also affect the biodegradation. Nonbiodegradable nanoparticles accumulate in the liver and spleen leading to potentially toxic side effects (Naahidi et al. 2013).

The nanoparticles used in paints, fabrics, cosmetics, and so on, can reach to the environment either intentionally or unintentionally resulting in potential source of environmental pollutions. Land and water surfaces will be polluted by the deposition of these wastes, which will consequently decrease the quality of these resources and disturbs the ecosystem. Nanoparticles in solid wastes, waste water effluents, spillages, or direct discharges can be transported to aquatic systems and this can have negative impact on aquatic life. The nanoparticles released in the air can easily pollute the air and if they are inhaled, acute and chronic respiratory diseases can be caused (Ray et al. 2009).

It has been found that the deposition of nanoparticles of 20 nm is 2.7 times higher than that of the particle of 100 nm and 4.3 times more than that of the particle of 200 nm (Stahlhofen et al. 1989). Kreyling et al. (2006) have reported

TABLE 9.3

Some Nanoparticles and Their Possible Health Risk

Nanomaterials	Possible Health Risks
Carbon nanotubes	Pulmonary inflammation, alveolitis, fibrosis, granulomas
Quantum dots	Cytotoxic to T-cell lymphoma
Silver nanoparticles	Argyria (blue–gray discoloration of skin and other organs) breathing problem, allergies, lung and throat irritation
TiO_2 and ZnO	Damage to DNA
Gold nanorods	Toxic to human skin
CuO	Cytotoxic and DNA damage

Source: Adapted from Ray, P.C. et al., *J. Environ. Sci. Health C Environ. Carcinog Ecotoxicol Rev.*, 27, 1–35, 2009.

that the nanoparticle deposition efficiency was higher in patients suffering from asthma or chronic obstructive pulmonary disease than in normal and healthy humans. This may be because of reduced clearance ability in those patients.

The toxic effects of some nanoparticles on the human health are summarized in Table 9.3.

9.5 Conclusion

Therefore, this chapter explains the principle and application of nanobiointerface and molecular recognition on biological science. The applications of nanoparticles are expanding day-by-day and it is necessary to understand the biocompatibility, biodegradability, and safety of these particles.

References

Birhanu, G., H. A. Javar, E. Seyedjafari, and A. Zandi-Karimi. 2017. Nanotechnology for delivery of gemcitabine to treat pancreatic cancer. *Biomedicine and Pharmacotherapy* 88:635–643.

Borel, T. and C.M. Sabliov. 2014. Nanodelivery of bioactive components for food applications: Types of delivery systems, properties, and their effect on ADME profiles and toxicity of nanoparticles. *Annual Review Food Science Technology* 5:197–213.

Bouwmeester, H., P. Brandhoff, H. J. P. Marvin, S. Weigel, and R. J. B. 2014. State of the safety assessment and current use of nanomaterials in food and food production. *Trends in Food Science and Technology* 40:200–210.

Buzea, C., I. I. Pacheco, and K. Robbie. 2007. Nanomaterials and nanoparticles: Sources and toxicity. *Biointerphases* 2:7–71.

Castner, D. G. and B.D. Ratner. 2002. Biomedical surface science: Foundations to frontiers. *Surface Science* 500:28–60.

Chan, W. C. W and S. Nie. 1998. Quantum dot bioconjugates for ultrasensitive nonisotopic detection. *Science* 281:2016–2018.

Chellaram, C., G. Murugaboopathi, A. A. John, R. Sivakumar, S. Ganesan, S. Krithika, and G. Priya. 2014. Significance of nanotechnology in food industry. *APCBEE Procedia* 8:109–113.

Couvreur, P. 2013. Nanoparticles in drug delivery: Past, present and future. *Advanced Drug Delivery Reviews* 65:21–23.

Dalkıran, B., P. E. Erden, and E. Kılıç. 2017. Amperometric biosensors based on carboxylated multiwalled carbon nanotubes-metal oxide nanoparticles-7,7,8,8-tetracyanoquinodimethane composite for the determination of xanthine. *Talanta* 167:286–295.

Daniels, T. R., T. Delgado, G. Helguera, and M. L. Penichet. 2006. The transferrin receptor part II: Targeted delivery of therapeutic agents into cancer cells. *Clinical Immunology* 121:159–176.

Desai, N. 2012. Challenges in development of nanoparticle-based therapeutics. *The American Association of Pharmaceutical Scientists Journal* 14:282–295.

Donahue, A. C. and M. Albitar. 2010. Antibodies in biosensing. In *Recognition Receptors in Biosensors* (Ed.) M. Zourob, pp. 221–248. New York: Springer.

Du, X., Y. Li, Y.-L. Xia, S.-Meng Ai, J. Liang, P. Sang, X.-L.Ji, and S-Q. Liu. 2016. Insights into protein–ligand interactions: Mechanisms, models, and methods. *International Journal of Molecular Sciences* 17:1–34.

Elghanian, R., J. J. Storhoff, R. C. Mucic, R. L. Letsinger, and C. A. Mirkin. 1997. Selective colorimetric detection of polynucleotides based on the distance-dependent optical properties of gold nanoparticles. *Science* 277:1078–1080.

Ezhilarasi, P. N., P. Karthik, N. Chhanwal, and C. Anandharamakrishnan. 2013. Nanoencapsulation techniques for food bioactive components: A review. *Food and Bioprocess Technology* 6:628–647.

Farah, A. A., R. Sukor, A. B. Fatimah, and S. Jinap. 2016. Application of nanomaterials in the development of biosensors for food safety and quality control. *International Food Research Journal* 23(5):1849–1856.

Fonseca, N. A., A. C. Gregório, A. Valério-Fernandes, S. Simões, and J. N. Moreira. 2014. Bridging cancer biology and the patients' needs with nanotechnology-based approaches. *Cancer Treatment Reviews* 40:626–635.

Garimella, R. and A. E. M. Eltorai. 2017. Nanotechnology in orthopedics. *Journal of Orthopaedics* 14:30–33.

Gomathi, P., D. Ragupathy, J. H. Choi, J. H. Yeum, S. C. Lee, J. C. Kim, and H. D. Ghim. 2011. Fabrication of novel chitosan nanofiber/gold nanoparticles composite towards improved performance for a cholesterol sensor. *Sensors and Actuators* 153:44–49.

Hannon, J. C., J. Kerry, M. Cruz-Romero, M. Morris, and E. Cummins. 2015. Advances and challenges for the use of engineered nanoparticles in food contact materials. *Trends in Food Science and Technology* 4:43–62.

Harmata V. and G. Náray-Szabó. 2009. Theoretical aspects of molecular recognition. *Croatica Chemica Acta* 82:277–282.

Heiligtag, F. J. and M. Niederberger. 2013. The fascinating world of nanoparticle research. *Materials Today* 16:262–271.

Hutter, E. and D. Maysinger. 2013. Gold-nanoparticle-based biosensors for detection of enzyme activity. *Trends in Pharmacological Sciences* 34:497–507.

Jahanshahi, M. and Z. Babaei. 2008. Protein nanoparticle: A unique system as drug delivery vehicles. *African Journal of Biotechnology* 7:4926–4934.

Joye, I. J. and D. J. McClements. 2013. Production of nanoparticles by anti-solvent precipitation for use in food systems. *Trends in Food Science and Technology* 34:109–123.

Konig, B. 1995. Molecular recognition. The principle and recent chemical examples. *Journal Fur Praktische Chemie* 33:339–346.

Kreyling, W. G., M. Semmler-Behnke, and W. Möller. 2006. Ultrafine particle-lung interactions: Does size matter? *Journal of Aerosol Medicine and Pulmonary Drug Delivery* 19:74–83.

Kwon, S. J. and A. J. Bard. 2012. DNA analysis by application of Pt nanoparticle electrochemical amplification with single label response. *Journal of the American Chemical Society* 134(26):10777–10779.

Lammerhofer, M. 2009. The world of molecular recognition. *Journal of Separation Science* 32:1489–1490.

Lehn, J. M. 1973. Design of organic complexing agents strategies towards properties. *Structure and Bonding* 16:1–70.

Leroueil, P.R., S. A. Berry, K. Duthie, G. Han, V. M. Rotello, D. Q. Mcnerny, J. R. Baker, B. G. Orr, and M. M. Banaszak Holl. 2008. Wide varieties of cationic nanoparticles induce defects in supported lipid bilayers. *Nano Letters* 8:420–424.

Li, S.D. and L. Huang. 2010. Stealth nanoparticles: High density but sheddable PEG is a key for tumor targeting. *Journal Controlled Release* 145:178–181.

Li, X., L. Wang, Y. Fan, Q. Feng, and F. Cui. 2012. Biocompatibility and toxicity of nanoparticles and nanotubes. *Journal of Nanomaterials* 2012:1–19.

Liu, J., J. Cai, H. Chen, S. Zhang, and J. Kong. 2016. A label-free impedimetric cytosensor based on galactosylated gold-nanoisland biointerface for the detection of liver cancer cells in whole blood. *Journal of Electrochemistry* 781:103–108.

Luo, X., A. Morrin, A. J. Killard, and M. R. Smyth. 2006. Application of nanoparticles in electrochemical sensors and biosensors. *Electroanalysis* 18:319–326.

Mahon, E., A. Salvati, F. B. Bombelli, I. Lynch, and K. A. Dawson. 2012. Designing the nanoparticle–Biomolecule interface for "targeting and therapeutic delivery." *Journal of Controlled Release* 161:164–174.

Markides, H., M. Rotherham, and A. J. El Haj. 2012. Biocompatibility and toxicity of magnetic nanoparticles in regenerative medicine. *Journal of Nanomaterials* 2012:1–11.

Mou, X., Z. Ali, S. Li, and N. He. 2015. Applications of magnetic nanoparticles in targeted drug delivery system. *Journal of Nanoscience and Nanotechnology* 15:54–62.

Mu, L. and S.S. Feng. 2003. A novel controlled release formulation for the anticancer drug paclitaxel (Taxol): PLGA nanoparticles containing vitamin E TPGS. *Journal of Controlled Release* 86:33–48.

Naahidi, S., M. Jafari, F. Edalat, K. Raymond, A. Khademhosseini, and P. Chen. 2013. Biocompatibility of engineered nanoparticles for drug delivery. *Journal of Controlled Release* 166:182–194.

Nel, A. E., L. Mädler, D. Velegol, T. Xia, E. M. V. Hoek, P. Somasundaran, F. Klaessig, V. Castranova, and M. Thompson. 2009. Understanding biophysicochemical interactions at the nano–bio interface. *Nature Materials* 8:543–557.

Nelson, D. L. and M. M. Cox. 2012. *Lehninger Principles of Biochemistry*. New York: W.H. Freeman.

Nie, S., Y. Xing, G. J. Kim, and J. W. Simons. 2007. Nanotechnology applications in cancer. *Annual Review of Biomedical Engineering* 9:257–288.

Ray, P. C., H. Yu, and P. P. Fu. 2009. Toxicity and environmental risks of nanomaterials: Challenges and future needs. *Journal of Environmental Science and Health Part C Environmental Carcinogenesis & Ecotoxicology Review* 27:1–35.

Shen, Z., Y. Li, K. Kohama, B. Oneill, and J. Bi. 2011. Improved drug targeting of cancer cells by utilizing actively targetable folic acid-conjugated albumin nanospheres. *Pharmacological Research* 63:51–58.

Stahlhofen, W., G. Rudolf, and A. C. James. 1989. Intercomparison of experimental regional aerosol deposition data. *Journal of Aerosol Medicine and Pulmonary Drug Delivery* 2:285–308.

Sun, A. L., Q. A. Qi, Z. L. Dong, and K. Z. Liang. 2008. An electrochemical enzyme immunoassay for aflatoxin B1 based on bio-electrocatalytic reaction with room-temperature ionic liquid and nanoparticle modified electrodes. *Sensing and Instrumentation for Food Quality and Safety* 2:43–50.

Tang, Y. H. and H. P. Zhang. 2016. Theoretical understanding of bio-interfaces/bio-surfaces by simulation: A mini review. *Biosurface and Biotribology* 2:151–161.

Tartaj, P., M. P. Morales, T. Gonzalez-Carreno, S. Veintemillas-Verdaguer, and C. J. Serna. 2005. Advances in magnetic nanoparticles for biotechnology applications. *Journal of Magnetism and Magnetic Materials* 290–291:28–34.

Thakor, A. S. and S. S. Gambhir. 2013. Nanooncology: The future of cancer diagnosis and therapy. *Cancer Journal for Clinicians* 63:395–418.

Thorstensen, K. and I. Romslo. 1990. The role of transferrin in the mechanism of cellular iron uptake. *Biochemical Journal* 271:1–9.

Walczyk, D., F. B. Bombelli, M. P. Monopoli, I. Lynch, and K. A. Dawson. 2010. What the cell "sees" in bionanoscience. *Journal of the American Chemical Society* 132:5761–5768.

Wang, C. and V. K. Yadavalli. 2014. Investigating biomolecular recognition at the cell surface using atomic force microscopy. *Micron* 60:5–17.

Wang, E. C. and A. Z. Wang. 2014. Nanoparticles and their applications in cell and molecular biology. *Integrative Biology* 6:9–26.

Wang, J., S. Tian, R. A. Petros, M. E. Napier, and J. M. DeSimone. 2010. The complex role of multivalency in nanoparticles targeting the transferrin receptor for cancer therapies. *Journal of the American Chemical Society* 132:11306–11313.

West, J. L. and N. J. Halas. 2000. Applications of nanotechnology to biotechnology. *Current Opinion in Biotechnology* 11:215–217.

Zeng, Y., Z. Zhu, D. Du, and Y. Lin. 2016. Nanomaterial-based electrochemical biosensors for food safety. *Journal of Electroanalytical Chemistry* 781:147–154.

Zhang, J., M. P. Landry, P. W. Barone et al. 2013. Molecular recognition using corona phase complexes made of synthetic polymers adsorbed on carbon nanotubes. *Nature Nanotechnology* 8:959–968.

Zhang, J., X. He, P. Zhang, Y. Ma, Y. Ding, Z. Wang, and Z. Zhang. 2015. Quantifying the dissolution of nanomaterials at the nano-bio interface. *Science China* 58:761–767.

Zheng, M., Z. G. Li, and X.Y. Huang. 2004. Ethylene glycol monolayer protected nanoparticles: Synthesis, characterization, and interactions with biological molecules. *Langmuir* 20:4226–4235.

Zhu, G. and H. J. Lee 2017. Electrochemical sandwich-type biosensors for α–1 anti-trypsin with carbon nanotubes and alkaline phosphatase labeled antibody-silver nanoparticles, *Biosensors and Bioelectronics* 89:959–963.

Index

Note: Page numbers followed by f and t refer to figures and tables respectively.